T0260971

CONNECTED SCIENCE

SCHOLARSHIP OF TEACHING AND LEARNING

Editors
Jennifer Meta Robinson
Whitney M. Schlegel
Mary Taylor Huber
Pat Hutchings

CONNECTED SCIENCE

Strategies for Integrative Learning in College

Edited by Tricia A. Ferrett, David R. Geelan, Whitney M. Schlegel, and Joanne L. Stewart

Foreword by Mary Taylor Huber and Pat Hutchings

Indiana University Press

Bloomington and Indianapolis

This book is a publication of

Indiana University Press
Office of Scholarly Publishing
Herman B Wells Library 350
1320 East 10th Street
Bloomington, Indiana 47405 USA

iupress.indiana.edu

| *Telephone orders* | 800–842–6796 |
| *Fax orders* | 812–855–7931 |

© 2013 by Indiana University Press

Manufactured in the United States of America

Library of Congress Cataloging-in-Publication Data

Connected science : strategies for integrative learning in college / edited by Tricia A. Ferrett, David R. Geelan, Whitney M. Schlegel, and Joanne L. Stewart ; Foreword by Mary Taylor Huber and Pat Hutchings.
       pages cm — (Scholarship of teaching and learning)
   Includes bibliographical references and index.
   ISBN 978-0-253-00927-2 (cloth : alk. paper)
   ISBN 978-0-253-00939-5 (pbk. : alk. paper)
   ISBN 978-0-253-00946-3 (ebook)
   1. Science—Study and teaching (Higher) 2. Interdisciplinary approach in education. I. Ferrett, Tricia A. II. Geelan, David. III. Schlegel, Whitney M. IV. Stewart, Joanne L.
   Q181.C595 2013
   507.1'1—dc23

                                        2013000402

1 2 3 4 5 17 16 15 14 13

Contents

Part IV. Broader Contexts for Integrative Learning

Foreword

The Scholarship of Integrative Teaching and Learning

Mary Taylor Huber and Pat Hutchings

THIS BOOK EMERGES from the intersection of two important reform initiatives in higher education. The first involves the growth of a scholarship of teaching and learning among faculty across disciplines, and the second concerns the support of integrative learning among undergraduates across their college careers.

The combination of these two developments can be powerful: faculty look closely and critically at classroom practice and student work in order to better understand and help students develop as integrative learners, able to connect their emerging knowledge, skills, and commitments across diverse settings. By asking questions about their students' learning, seeking evidence to answer those questions, using that information to improve instruction, and engaging colleagues with what they are finding, these faculty are not only creating better learning experiences in their classrooms and programs but also contributing to knowledge and field building around what it means to teach with "integrative learning" in mind.

This focus is timely, because integrative learning has become central to the effort at colleges and universities in the United States and beyond to rethink and redesign liberal education for the 21st century. While educators have long endorsed the value of integration, the burden has traditionally fallen on the learner, with campuses assuming that bright students would be able to pull together the pieces of their education on their own. Recent thinking about liberal education has taken a different stance. What's new is the conviction that institutions should make this a goal for all students, and do what they can through the curriculum, cocurriculum, pedagogy, and assessment to help them realize the importance of integration and to have multiple opportunities to practice and perfect the needed skills (Huber and Hutchings, 2005).

The needs and opportunities are, perhaps, especially great in the science fields. Many of the 21st century's most pressing challenges—climate change, energy policy, food and water sufficiency, public health, medicine, information security—have strong science components that span multiple disciplines. Higher education must attract and graduate a larger and more diverse group of professionals in science; it must develop their capacities for interdisciplinary synthesis and cross-disciplinary collaboration; and—beyond those professional scientists, technologists, engineers, and mathematicians—it must find ways to create a more scientifically literate general populace. The fruits of such efforts go both ways, certainly, to society through better science and more informed civic participation, and to students who, through grappling with meaningful issues, come to a fuller understanding of the underlying science as well as a greater appreciation for the power of integrative thinking in its many varieties and forms.

Indeed, as Tricia Ferrett points out in her introductory chapter, integrative learning has long been key to movements for science education reform. Dismayed by curricular overspecialization, educators have sought to restore science to the liberal arts, linking it to "the human domain." Alarmed by low levels of science literacy, professors have called for interdisciplinary approaches to college science focused on public issues rich in science content; hoping to encourage participation by underrepresented groups, reformers have experimented with more active, engaged pedagogies; concerned about the rapid growth of knowledge and information, innovators are foregrounding scientific ways of knowing, seeing content coverage as an ever more elusive goal.

The term "connected science" picks up the common, integrative thread running through these movements, and adds to it what we might call a theory of pedagogical action—namely that faculty should bring their skills as scientists and scholars to key questions as they arise in their own classrooms and programs. The integration of learning is widely considered to be a relatively sophisticated skill, which develops over time and requires considerable effort and experience to attain. What are its varieties? What are the developmental trajectories for these kinds of thinking? How can faculty help students hone more sophisticated skills for integration? How do students learn to evaluate the connections between concepts, courses, and contexts that they and others make? Structures that transcend, complement, cut across, or link college programs and courses are important, but so too is the pedagogy that lies at the very heart of these programs and courses themselves.

This is territory that calls for the scholarship of teaching and learning. As evidenced by the accounts in this volume, such work emerges from and reflects the thoughtfulness with which faculty construct the learning environments they offer students, the attention they pay to students and their learning, and the engagement they seek with colleagues on all things pertaining to education in their disciplines, programs, and institutions. Scholars of teaching and learning use a variety of pedagogies and pursue a variety of learning goals, but most are open to, indeed seek, new ways to help their students reach more challenging educational goals. This is certainly so for those, like the contributors to *Connected Science*, who have studied students' integrative strategies, designed new learning experiences, and experimented with a variety of teaching approaches, in order to help their students strengthen their integrative learning skills.

First on the agenda, for several authors, is finding out what kinds of moves their students are using to bridge disparate subject matters (like science and philosophy) or contexts (a science topic and a policy issue). In the second of her two introductory essays, Ferrett suggests that "part of the difficulty in articulating student learning outcomes for connected science is that as instructors, we naturally start from a position of expertise in our fields"—a position that can make one blind to and mute about the moves that novice learners need to master. She notes, too, how little is known about the intellectual development of students' integrative and innovative capacities during college, and calls attention to what many teachers (and not just in the sciences) have observed: "students, for their most fruitful integrative work, will likely draw on the course material and their own personal perspectives from prior knowledge, experience, and belief."

In order to better understand (and strengthen) beginning students' integrative skills, Ferrett (chemistry, Carleton College) and Joanne Stewart (chemistry, Hope College) both taught freshman science seminars designed to elicit integrative thought. Both gave an assignment asking students to make analogies and to name something new they'd learned from doing so—an exercise that enabled these two teacher-scholars to devise a typology of students' integrative "moves" (as they discuss in their coauthored essay in this collection), and also underlined for students the metacognitive lesson: that such analogies can be created and explored.

Bettie Higgs (geology, University College Cork, in Ireland) redesigned her field course from the standard "lecture in the field" format to a "seminar in the field," where teams of students record their observations in ways that give them opportunities to connect what they're seeing with what they're learning in other courses, as well as to experience how campus-based science connects with phenomena in the real world. Higgs's studies of student performance in the field informed the development of new assignments that buttress students' abilities to make these kinds of connections. Her conclusion from such work is an important one: that integrative learning doesn't just happen. It's not inevitable or "straightforward," but needs what she calls (borrowing from astrophysicists) "exotic matter" to happen. And while the teacher may initially provide the necessary "irritant" for this kind of learning, the goal is for students to eventually provide it themselves.

The sciences offer students special irritants for such connection making, of course: the challenge is to make productive use of them, so that students don't simply shut down. Mike Burke (mathematics, College of San Mateo), teaching quantitative literacy to students in mathematics, uses data-based writing assignments focused on practical problems, like global warming, in his classes. He finds that students often have trouble "discarding their preconceptions and deciding precisely what conclusions are truly supported by their work with data." Yet because using data to make decisions is such an important skill, Burke believes that "it is our job as teachers to create this kind of discomfort," and to help students move through it, becoming better, more integrative, problem solvers in the process.

Designing learning experiences that invite integration is critical to this task. Xian Liu (English), Kate Maiolatesi (environmental sciences), and Jack Mino (psychology), all at Holyoke Community College, took several years to develop their interdisciplinary learning community course to help students connect science and writing in "contextualized" and "problem-centered" ways. Focused on sustainability, the authors engaged their students in field-based projects to explore sustainability principles in practice, and communicate results in writing to community partners. Matt Fisher (chemistry, Saint Vincent's College), too, chose practical problems as a vehicle for encouraging integrative thinking in a sequence of courses for biochemistry majors. Teaching "through" public health issues to the "underlying basic biochemical concepts" enabled his students to match the levels of understanding achieved by those in more conventionally taught programs, but also to become "more engaged affectively with the course material" and more attuned to the moral dimensions of scientists' roles in society.

As Fisher's work suggests, designing for integrative learning often means involving students more authentically in activities that capture aspects of real-world science. Gregory Kremer (mechanical engineering, Ohio University) reports on the effectiveness of team-based "connected capstone" projects designed in response to actual community needs, in which students develop the technical, professional, and social skills needed to *do* engineering but also, importantly, to *be* engineers. Robert Brooker, David Matthes, Robin Wright, Deanna Wassenberger, Susan Wick, and Brett Couch (from various biology fields, University of Minnesota, Twin Cities) describe their institution's transformation of the introductory biology course from a lecture format to team-based learning that makes extensive use of instructional technologies. Based on the widely admired SCALE-UP model, the aim is to "introduce . . . students to the real work of being a biologist in their first major college course in the discipline." Though challenging both for faculty and for students accustomed to a more linear classroom experience, preliminary results suggest that students who complete this course earn significantly better grades in subsequent traditional-format biology courses and are retained in biology at a higher rate.

Teaching for integrative learning often helps faculty members develop a keener sense of the whole student, and an appreciation of the ways that knowledge, skills, and attitudes are interconnected—and may sometimes work at cross-purposes. David R. Geelan (science education, Griffith University), attentive to the knowledge and skills that future middle-school teachers will need to teach science, is also concerned with these students' "confidence and self-efficacy." How to develop a "teacher identity" that integrates personal and professional dimensions is key to the integrative pedagogies he has experimented with in his course. However, as Richard Gale (theater, Mount Royal University) points out, experimenting with pedagogies that target integrative learning need not mean inventing approaches that are entirely new: "Many of the strategies we know and love can be made to serve the needs of integration and improve the connective capacities of students; the key is not to revise so much as to reframe." Indeed, the essays of *Connected Science* give vivid testimony to the range of approaches that can be successfully employed toward these ends—and to the importance of honing and adapting them for particular contexts, circumstances, and purposes.

As integrative learning becomes more widely valued as a teachable outcome for liberal education, it is likely to provoke increasing numbers of faculty to ask how they, too, might do more for their own students. But the appeal of the idea is also a challenge. As Ferrett points out, integrative learning means teaching toward goals that are "highly related, rich, and difficult to articulate" and it thus "begs for a scholarship of teaching and learning" to address those goals with "a range of methodologies, questions, theories, and perspectives." Of course, there's room for the kind of research that the learning sciences uniquely provide. But there's also a need for cases, examples, stories, and studies by teachers who have struggled in their own classrooms with the pedagogical and methodological issues that the scholarship of *integrative* teaching and learning entails.

And it *is* sometimes a struggle. Courses and programs like the ones featured in this volume are born out of hard work, hard thinking, and resilience. In the final essay in the volume, Whitney Schlegel (biology, Indiana University Bloomington) recounts the very intentional process she and her colleagues undertook in creating her campus's interdisciplinary Human Biology Program. Central to this long-term effort was bringing people on board at different levels and stages all along the way—from state leaders to campus administrators, from faculty to students—in order to develop, implement, and sustain a common vision.

This is not the kind of thing that happens overnight. As members of the leadership team of the Carnegie Foundation's 12-year program on the scholarship of teaching and learning (CASTL—the Carnegie Academy for the Scholarship of Teaching and Learning), we have seen firsthand how the kinds of studies and reports collected in these pages emerge over time, beginning, often, in inchoate concerns and questions about student learning, taking shape in the design and selection of methods for the study, and moving through a cycle of visions and revisions as findings are shared with colleagues, critiqued, sharpened, and polished for presentation in public forms and forums.

What's particularly inspiring to us as observers and advocates of this work is seeing how one effort leads to another. Several of the projects reported in this volume had their genesis in CASTL, but as they took shape they attracted the interest and engagement of a wider circle of colleagues, as well—educators from diverse settings who saw the power of the scholarship of teaching and learning to shape new pedagogical practices and curricular designs for integrative learning. The result is the rich mix of contexts, disciplines, designs, methods, and insights that readers will find and benefit from in *Connected Science*. The essays have much to offer in their own right, but we know that they will catalyze further efforts as well, as commitments to integrative learning take fuller hold within the larger teaching commons where new ideas are tried out, traded and built upon, and, yes, integrated into future educational thought and practice in ways that will enrich the experience of students and teachers alike.

Reference

Huber, M.T. and Hutchings, P. (2005). *Integrative Learning: Mapping the Terrain*. Washington, DC: Association of American Colleges and Universities.

PART I

CONNECTED SCIENCE

Why Integrative Learning Is Vital

1 Fostering Integrative Capacities for the 21st Century

Tricia A. Ferrett

I OPEN WITH TWO stories to help frame the purpose and contributions of this book. These stories will provide concrete anchors for a more extended discussion of an approach to undergraduate science education—"connected science" learning and teaching.

Alice's Senior Biochemistry Thesis

Several years ago a student from Africa did her senior thesis on the design of new drugs for HIV AIDS. Alice had a strong biochemistry background, and she was drawn to the moral purpose of her topic. As she evaluated the pros and cons of first- and second-generation drugs, she learned not only about research on chemical structure and function relationships at the molecular level but about side effects and drug effectiveness in the human body. It was clear—human issues, not just scientific ones, were guiding research in the science of AIDS. Alice worked mostly alone, with my guidance as a chemistry instructor. As instructed, she became immersed in the scientific research literature and began to integrate her prior learning of chemistry. But while she was drawn to the human context, she was never entirely at ease with bringing it into her thesis. She had been explicitly asked to "do the chemistry deeply." One day Alice said, "what about if I do just a little bit of context in the introduction?" Alice was good at reading faculty signals; she knew to keep the human stuff off to the side. When I suggested she steer her paper and conclusions in a creative and synthetic direction that grabbed her, she was tentative at first. What would that mean? Would she be sacrificing the science in doing so? Is that allowable for the senior thesis? In the end,

Alice chose to propose a specific next-generation drug that overcame some difficulties encountered in earlier versions. Once she hooked onto this approach, she blossomed with a larger purpose to her work. Her motivation and creativity rose. Her scientific thinking was strongest here. Alice stepped over the threshold to create something that was uniquely hers—the structure and rationale for a new HIV drug.

Jeff's First-Year Study of Sustainability

Jeff was a first-year student in a learning community facilitated by Xian Liu and Kate Maiolatesi that integrated first-semester English language and literature with an introduction to sustainability studies. In teaching an honors course integrated across disciplines, the two instructors were committed to creating an atmosphere of community. They began with a kayak trip on a local river, where students were introduced to the concepts of complex ecological systems, aquatic ecology, and each other. In the trip, Jeff and his classmates began to develop a sense of each other's needs while engaging in the science. As the course progressed, Xian and Kate gave the students the option of doing a community-based project at one of two local sites—the community food bank's vegetable farm or an alternative high school. Half the students chose to work on the farm, which donates healthy produce to a local food bank used by low-income families. Other students worked as consultants for the high school director, researching how to make the campus green, the energy renewable, and the lunches healthy. Both options combined science, sustainability, community service, and links to social justice. At the end of the semester, Jeff believed that the learning community worked so well, in part, because students quickly became friends and spent deep time together learning in part through real-world experiences. Once Jeff came to understand the complexity of issues around sustainable living and the scientific concepts underlying personal choices, he wanted to take this to the next level. Fortunately, his college was preparing to offer a formal sustainability studies program. His excitement built as he discussed his next steps with Kate and Xian. Jeff proposed the development of a campus "sustainability center." The instructors agreed to have a cohort of students design a two-room green building to house the program and its classroom while linking to the community through demonstration projects and an on-site organic farm. Their vision also included teaching some introductory science labs at the farm. The green building would be accessible to those outside science and the program. The instructors and students imagined a busy, cool place to hang out, work, connect, and learn—a "science in action" community place.

What do these stories have to teach us about the promise and practice of college science learning, and its role in preparing students to live, work, lead, and learn in the complex and changing world of the 21st century? The first story departs subtly from traditions for college science teaching in order to move toward a more connected science. A senior worked on an "integrative" capstone exercise—integrative within the discipline, that is. Working alone, she drew from the original scientific literature, approaching the science with a critical eye while integrating and applying chemistry she had learned. Yet she was unpracticed with regard to letting a larger purpose steer her science learning. The integrative nature of her topic was notably understated in the final product. Her hesitation to include "context" shows the barriers to integration that instructors create when we project compartmentalized disciplinary norms onto our curriculum and students. Admirably, Alice displayed a deep engagement with her science and took her learning to a new level through the creative move of application.

The second story stretches a bit further toward a *vision of the possible* for science learning and teaching—a vision that connects science learning in more concrete and intentional ways to human issues. In Jeff's story, nature itself integrates issues related to sustainable futures first. On the human side, science informs our choices, but human nature comes into play when making those choices, as do political and economic needs. A sustainable lifestyle depends on the laws of conservation, recycling matter and energy, biodiversity, adaptation, population dynamics, and carrying capacity. From the beginning, students in the learning community are given the permission and support to connect, explore, and guide this work with purpose. The fact that their work matters to someone else produces higher student engagement, motivation, and commitment. Furthermore, this learning community has dismantled not only the intellectual boundaries between disciplines but physical boundaries as well. The classroom has become more porous and linked to the community. The learning community is critical in producing an environment in which the students' engagement, scientific understanding, and integrative capacities grow individually and together, over the semester. This book articulates scholarly evidence of student learning for a more coherent approach to undergraduate science education, which we call connected science learning and teaching. This approach borrows from, builds on, and synthesizes elements from prior and existing science reform movements and projects while articulating a unique educational philosophy that emphasizes the building of integrative capacities in our students. We show, very concretely, how the elements of connected science come together in various contexts and settings, and how a more systematic and scholarly examination of how this happens and with what outcomes can strengthen work in these directions.

Why Connected Science?

Why is connected science important in higher education today? The need to prepare students to engage with complex problems facing our global society in the 21st century is argued eloquently, with regard to general education ideals, by the Essential Learning Outcomes from the Liberal Education and America's Promise campaign of the Association of American Colleges and Universities (AAC&U, 2007). These general education ideals articulate four categories of learning outcomes: knowledge of human cultures and the physical and natural world, intellectual and practical skills, personal and social responsibility, and integrative learning. One of these categories, integrative learning, involves "synthesis and advanced accomplishment across generalized and specialized fields" (p. 3). Across the three other categories, there is also a persistent emphasis on engaging the big questions, both contemporary and enduring, for local and global communities through a focus on projects, problems, and issues. Connected science fits naturally into this larger framework for a 21st-century liberal education for college students. Furthermore, our students must "not only interpret the world, but take up a place within it as citizens, at work, and as whole persons," as is argued in *A New Agenda for Higher Education: Shaping a Life of the Mind for Practice* (Sullivan and Rosin, 2008), from the Carnegie Foundation. This "requires teaching for practical reasoning, a long tradition that has been overshadowed by the advance of specialized theory and abstract analysis," say William Sullivan and Matthew Rosin. The book discusses an engineering course in which students grapple with the perspectives of engineers in other cultures and a human-biology course that deals with the science and ethics of death. We concur with the general argument for a stronger marriage between the abstract and the practical in higher education. These engineering and human biology examples qualify as connected science.

This is an opportune time for connected science. For science educators, preparing our students to engage with complex problems by "practicing" analysis and action in the real world is critical at this point in earth and human history. As students like Jeff and Alice confront the science-rich issues of climate change, disease, and sustainability, there are overwhelming reasons to connect their science learning to human experience and "practical reasoning." This does not mean compromising on the rigor of the science or the depth of students' understandings about the natural and physical world. It also does not mean stepping away from the impressive standards for objectivity, process, and evidence that science has developed over the last few centuries. It does mean that we have a chance to further engage student interest and motivation to learn, drawing on their and our passions, experiences, and aspirations. Connected science will allow us to learn science together with our students, applied to things that matter in a larger sense. We can more often choose to learn science "for something"—in service of a cause—so students gain concrete experience in dealing with difficult multidimensional problems. Connected science also aims to base student learning of science on the science of human learning, make use of interdisciplinary and integrative content and pedagogies, and build programs that support in-depth approaches over time. I will elaborate below on integrative learning, its relationship to connected science, and more specific student learning goals for connected science. We, the authors of this volume, want to help students learn science knowledge and processes—and to practice complex analysis and sometimes act on this analysis in the world around them. As teachers, we don't mean to thin out the science learning, but rather to deepen and add more texture through integration, application, practice, and action.

Historical Foundations for Connected Science

Aspects of connected science teaching and learning at the college level are not entirely new, in aspiration or in practice. In my own life as a scientist and teacher, I have developed a strong attachment to the language of former Carleton College president and Antarctic explorer Larry Gould (1945): "[T]he true spirit of liberal or humane studies is not inherent in any special or sacred field. There are quite as great cultural values to be derived from the study of chemistry or geology as from that of Latin or Greek, if inspired teaching guides the students" (np). Gould's leadership gave the sciences a place at Carleton as a liberal art. Gould helped our college begin a move from "science and the liberal arts" to "science and the other humanities." This move linked science to the human domain, on more even footing with academic disciplines that are more traditionally connected to the study of human endeavors.

Several decades later, "issues" courses at liberal arts colleges sprang from the 1960s call for "relevance" in higher education (Hudes and Moriber, 1971). More than 40 years ago, Isidore Hudes and George Moriber wrote eloquently about the need to "make young people aware of . . . problems faced by everyone in society . . . by developing a course around those areas which are expected to dominate mankind for the next decade and beyond" (p. 162). Even in 1971, these authors were calling for college science courses that were interdisciplinary in nature and centered on environmental pollution, conservation, population control, and ecology. By 1989, Project 2061 of the American Association for the Advancement of Science (AAAS, 1989) argued that science literacy was crucial for all citizens who live in a world increasingly filled with science and technology and an array of problems facing humanity. This call to connect science learning to compelling issues of societal concern came well before the world became as connected and globalized as it is today. The "science for all" movement led to courses for majors and nonmajors that helped students become conversant with public issues with a significant science context.

With the more recent focus on engaging students from underrepresented groups in science, technology, engineering, and math (STEM) fields, the "science for all" notion has expanded to include a much-needed diversity dimension (Seymour, 2002). Several US reports articulated visions for this movement (NSF, 1996; NRC, 1996, 1999), including a call to discover which teaching techniques were most effective in engaging a more diverse set of learners. These issues of social justice deeply motivate a number of STEM reform movements, including our work in connected science.

This sense that undergraduate science education must respond more directly to the needs and problems of humanity has only grown in the last 20 years (Hake, 2000). For example, numerous reports from educators and scientists in the United States call for more interdisciplinary teaching and learning in the undergraduate science curriculum (PKAL, 2002). Again, the argument is that we need to prepare students, as citizens and scientists, to address the local and global problems of the 21st century. For almost two decades, communities of innovation have grown and developed around the development of teaching science "in context" or with pedagogies that more closely mimic authentic scientific inquiry and active learning. Groups like SENCER (Science Education for New Civic Engagements and Responsibilities), ChemConnections, Bio-QUEST, Project Kaleidoscope (PKAL), and the Howard Hughes Medical Institute (HHMI)—all of which have created or supported projects that teach science in a real-world context relevant to students—have contributed to substantial progress in the areas of faculty development and curriculum design. In addition, a range of "pedagogies of engagement" in science (Mestre, 2005) are being used and studied: guided inquiry (POGIL, 2013), problem-based learning (PBL, 2013), learning communities (LC National Center, 2013), team-based learning (TBL, 2013; Michaelsen et al., 2004), peer-led team learning (PLTL, 2013), case studies (National Center for Case Study Teaching in Science, 2013), and others. A recent article does a nice job of comparing three of these "pedagogies of engagement in science"—PBL, POGIL, and PLTL—and synthesizes results in a format useful for instructors who are making pedagogical decisions in their own contexts (Eberlein et al., 2008). Finally, PKAL's "Pedagogies of Engagement" project (Narum, 2008; PKAL, 2008) has worked with "pedagogical pioneers" in STEM fields in order to design professional development opportunities for existing networks of faculty and web resources that disseminate and synthesize reform lessons—from a pedagogical perspective, and has established a partnership with the Science Education Resource Center (SERC) to facilitate delivery of resources for faculty (SERC, 2013).

Interestingly, only a few of these major projects have significantly challenged the traditional content of science teaching. Teacher concern with the issue of "coverage" is a formidable barrier to the reforms that call for science learning in context and with more engaged pedagogies. Even the narrower issue of how to balance the teaching of "science content and process" can send college science instructors into a tizzy. These tensions have made the pace of science reform slower than it might have been in less established disciplines like political science or sociology, where the norm is to expose students to a full set of evolving and conflicting theories in the context of modern or enduring issues. The sciences justifiably take great pride in the large knowledge base and methodologies that have been developed; the degree of consensus in these disciplines is relatively high with regard to methodology (the scientific method) (Donald, 2002). Science instructors are quite serious about sorting out the "big ideas" in their fields as they prioritize and emphasize course content. However, the explosion of information and knowledge in the sciences and elsewhere, driven by revolutions in technology and biology, make the goal of "broad" coverage increasingly unrealistic for today's students. Related to this concern, biologist Craig Nelson has eloquently discussed what he thinks is a false dichotomy between teaching science content and process (Nelson, 1989). He argues that students who understand and engage in the processes of science more deeply are better situated to understand the scientific ideas. Nelson thus argues for a both-and approach to what he sees as the "mythical" dilemma between content and process. Wherever one stands with regard to this dilemma, similar challenges apply to connected science, further amplifying the "coverage problem" perceived by many science educators. Furthermore, none of the major reform efforts have argued clearly enough for a coherent approach that intentionally aligns science context, content, and pedagogy around a well-articulated educational philosophy. In an article that tracks the processes of change in US STEM education in higher education (Seymour, 2002), Seymour notes that "[l]earning is enhanced when all the main elements in a class fit coherently and overtly together: class content and activities, lab work, assignments, the text, media, and other resources" (p. 96). Seymour goes on to note the complexity of the human social system surrounding learning, and puts forth a theory: "Attempts to alter single elements in a complex social system will not be effective; each element must be aligned with the others for system changes to prevail" (p. 96). We believe this notion of alignment is a theory well worth testing as reform in undergraduate education moves forward. The next stage of reform must not only address the changing world of the 21st century for our students, it must also attempt to link elements of courses and programs in a coherent way. Connected science teaching and learning is, in part, an expression of both of these aspirations.

The Role of Assessment and Scholarship

Turning now to hard questions about what we really know about "what has worked and why" for students in the history of reform in undergraduate science education, I look first to the realm of educational assessment and project evaluation. Impressively, assessment work tied to these efforts has multiplied; we heartily endorse the continuity of this work. Much of the work rests on surveys where students self-report on their experiences and their learning. However, locating more in-depth research grounded in the analysis of actual student work is difficult and often context specific. Where assessment results on student learning do exist, they tend to be thin, dispersed, and difficult to access; some of the results exist, for example, only in project or evaluation reports. Synthesizing the results of projects and contexts is even more difficult. Generally, available results align with those from the science of learning. For example, students need to be actively engaged, they need to construct understanding in light of their prior knowledge and beliefs, and deeper learning is recursive and contextualized. In 2004, an article in *Science* discussed the notion of "scientific teaching" (Handelsman et al., 2004), calling for educators and others to apply the same level of rigor to teaching as we do to scientific research. In addition, these authors assert that "Scientific teaching involves active learning strategies to engage students in the process of science and teaching methods that have been systematically tested and shown to reach diverse students" (p. 521). DeHaan (2005) has nicely summarized some of the evidence to support active learning strategies, though not yet in a way that is linguistically appropriate for most STEM educators. Wieman (2007) succeeded in reaching a larger audience with his article on a scientific approach to science teaching and how research results can be applied to classroom teaching. Prince and Felder (2007) describe a range of evidence-based science teaching methods along with issues of implementation. Nonetheless, the upshot is that despite the most recent attempts at distilling the results of careful research, it is hard to study student learning rigorously and equally hard for STEM faculty to locate results that are accessible and robust in a scholarly sense. For the subset of results relevant to the approach of connected science, these problems are compounded even further. Thus, there is a pressing need for *integrated* and *collective* scholarship practiced by expert teachers and scholars from a range of scientific fields and spanning multiple contexts. To this end, our international author group of STEM instructors, supported by the Carnegie Foundation for the Advancement of Teaching and its complementary Integrative Learning Project with AAC&U (Huber et al., 2007), has focused on connected science within and across courses, programs, and institutions. Most of the authors first met each other during the yearlong 2005–2006 Carnegie Scholars program around the theme of integrative learning. Through our project work with senior scholars at the Carnegie Foundation, we have learned to complement our disciplinary expertise with methods for investigating student learning through a "scholarship of teaching and learning." Scholarship like this has the capacity to address larger questions across a range of contexts and projects—through the lens of scientific experts who are vitally present in the classrooms where we design, teach, and study.

This book attempts to move existing work in assessment and scholarship to a new level, by coordinating and synthesizing research by multiple scholars who are tackling the nuances and complexities of connected science in real and varied classroom settings. Our collective research circles around this rich cluster of questions—what does connected science learning and teaching really look like, how might it look at its best, how does it work, and why? The case studies that the following chapters comprise put forth a set of diverse models from many fields of science, math, and engineering—along with scholarly evidence and synthesized insight. Through their case studies, these scholars encounter and study the expected, and quite often the unexpected.

Finally, we take note of the assessment climate in the United States and abroad, where a chasm is growing between the skills and knowledge valued in liberal learning and science learning and the things being measured, mostly in standardized ways. At the same time, much good work in assessment is aligned with institutional goals for student learning and recognizes the need to use methods that are grounded in more nuanced analysis of student work. Like others, we have started to address student learning goals that will be of high value as our students enter a connected, unstable, and global world. We also realize that outcomes related to these goals are hard to measure; they can easily be missed in standardized and traditional classroom testing. Our scholarship provides ideas for how instructors can naturally embed into course assignments and activities assessments for some of the learning goals tied to connected science, with examples.

We hope this book, a modest beginning, will help educators think about, tweak, and develop curricula using an evidence-based approach, starting from where they are in their relationship to connected science. In a practical sense, the scholarship in this book will illustrate a range of rich and generative ways to think about, design for, and assess science learning where the development of integrative capacities applied to real-world problems is an important goal. By example, we also aim to inspire others who hold a deep interest in student learning to integrate this kind of scholarship into their professional lives.

The Case Studies in This Book

Connected science is brought to life through the cases studies in this book. Some essays focus on connections between science learning and societal issues while others aim to have students integrate their science learning with professional functions or within a single discipline. Other authors study how teachers help students create something "new" through the integration of one or more perspectives or disciplines. Still others are studying the processes in which instructors engage as they create integrative programs. This group of scholarly case studies is a collection of *multiple voices*. Dogma, an exhaustive approach, complete coverage of the field—these are not our aims.

The case studies start with a group of essays that focus on student learning at the single-course level, yet through varied lenses. Matt Fisher writes about courses for science or engineering majors. He faces the quite thorny and ubiquitous "coverage problem": he expects his students to learn deeply in the discipline while connecting this learning to real-world contexts. Fisher uses issues of public health—HIV AIDS, alcohol abuse, bird flu—to help his students connect their science learning in a biochemistry course to personal and institutional values. At his Catholic college, this is an approach strongly aligned with institutional mission. Fisher's evidence aligns with Nelson's argument above, though for Fisher the "coverage problem" involves a focus on values rather than science process.

Gregory Kremer at Ohio University considers ways to educate engineers with a broad vision of their profession and the key integrative skills required of them. He outlines the development and teaching of a capstone course in the engineering program designed to take students beyond disciplinary knowledge to consider rich real-world engineering challenges.

Mike Burke talks about "scientific thinking" as his introductory math students apply and write about quantitative modeling related to problems like global warming and the Irish potato famine. He asks his students to argue from evidence, not backward from preconceived conclusions. His work is aligned with efforts by many others to expand the role of quantitative reasoning across the college curriculum. The strategy of writing with mathematics and scientific data and thinking is an integrative teaching approach from which we can learn. Burke's work also raises fundamental questions about the developmental challenges of teaching young college students to think scientifically about complex issues.

Bettie Higgs studies the renovation of a first-year science residential field course. She aims to help students make explicit connections to prior knowledge in science and to the natural world. Through the addition and study of a number of assignments, she develops a more coherent pedagogy for integrative learning. Conceptually, she introduces and expands on the metaphor of a "wormhole," a pathway that helps students connect and transform their learning. This metaphor helps Higgs more closely examine critical moments in integrative learning, and the difficulties some students still encounter when opportunities to connect learning are provided.

The remaining parts of the book move to yet broader contexts: K–12 education, institutions, learning communities, and programs. David R. Geelan tells the story of a course that helps future middle-school teachers connect the teaching of integrative science to literacy learning while they simultaneously focus on their development as teachers. His chapter explores the implications of this approach for helping teacher education students with limited science backgrounds deliver high-quality, integrated science teaching to their students. Geelan has created a "pedagogy of integration" that helps students develop their identities as professional teachers.

Robert Brooker, David Matthes, Robin Wright, Deena Wassenberg, Susan Wick, and Brett Couch at the University of Minnesota offer the example of revising teaching in large undergraduate courses in introductory biology to enhance students' development of skills in biology as well as a broader approach to knowledge. They demonstrate that it is possible to attend to integrative learning even within the context of courses that are unavoidably taught in large lecture format.

Xian Liu, Kate Maiolatesi, and Jack Mino examine change over a much longer time span by reporting on research about student learning undertaken over five years in a first-term honors learning community on sustainability. Indeed, this is the course our friend Jeff inhabited at the beginning of this introduction. Their teaching extends the furthest past the traditional classroom, involving field-based labs and community projects.

Likewise, Richard Gale creatively synthesizes a number of ideas in his chapter on integrative pedagogies. He entices us to think in novel ways, for the purposes of integrative learning, about a range of pedagogical friends—ones in new places, old ones in new clothes, and more—and "signature pedagogies" (Shulman, 2005). His broad range of exemplars, spanning teaching contexts inside and well outside the sciences, paints a colorful and inspiring portrait of instructors who are applying pedagogy with intention.

In the final part of the book, the chapter I coauthor with Joanne Stewart (Hope College) investigates the integrative "moves" that students make in a data-rich course on abrupt climate change. These courses were taught at two very different colleges with very different student populations. The work, centered on understanding the creative act of integration itself, reflects on how the same approach plays out in different institutional contexts.

Closing out the book, Whitney Schlegel—founding director of the Human Biology Program at Indiana University—focuses on a scholarly approach to the instructor development that underpinned the multiyear creation of this new integrative program. Her chapter discusses development of this integrative curriculum and examines how this process has inspired change across her campus.

A Connected Science Signature?

If we play the stories of Alice and Jeff forward now in a fictional way, what insights about connected science learning might they give back to us? Let's imagine that Alice's interest in HIV AIDS leads her to a graduate program where she is doing interdisciplinary research on AIDS. She attends a Boston conference on biotechnology and sustainability, where she meets Jeff through a common friend. He is attending his first professional meeting with his sustainability studies program from his community college. They talk one evening over coffee about their current and past educational experiences. Jeff talks about how the web of interconnectedness he is learning about in natural ecosystems seems to be mirrored by the web of human relationships in his interdisciplinary program. He can barely keep track of it all, but he feels like he is thriving. Alice agrees, noting that she is only beginning to understand the range of issues relevant to her research on AIDS. Culture, economics, issues at the level of a human individual or molecule. Not to mention the challenges of developing treatment strategies that work not only in the human body but in all these other contexts as well. She is swimming, at the same time, in a sea of conferences and new colleagues Each human interaction adds to and connects her learning in a new and powerful way. Together they come to realize that the "learning systems" for college science students could benefit from being more integrated, and connected—just like the systems they both study in the natural world and experience in their professional lives.

It is probably too early to articulate a well-defined signature for connected science. However, we are tempted to claim that the signature resides somewhere inside this conversation between Jeff and Alice. Human learning takes place, in part, in the human brain, itself a highly networked complex system. The complexity of human learning also extends into the networked social and knowledge systems that support our students. How could learning and teaching at its best be conceptualized as anything but complex, interconnected, nonlinear, and integrative?

Connected science, with its emerging emphasis on bringing coherence to a set of design elements for integrative learning—content, context, and pedagogy—is a comprehensive, applied philosophy. Though complex in nature, we believe it holds promise as an educational compass for those who wish to prepare college students for 21st-century encounters with science.

Acknowledgments

Chapters 1 and 2 have benefited from the fresh ideas and perspectives of authors of this volume. Kate Maiolatesi and Jack Mino provided the story about Jeff, a composite student from Holyoke Community College, where they both teach. Joanne Stewart, Matt Fisher, Bettie Higgs, Whitney Schlegel, Mary Huber, and Pat Hutchings helped to edit and critique earlier drafts of this chapter. Sandra Laursen (University of Colorado Boulder, ethnographic and evaluation research) provided essential "outsider" critique and comment as a veteran science educator and reformer who also specializes in assessment and evaluation of varied science education projects from K–12 through higher education.

References

AAAS (American Association for the Advancement of Science). (1989). *Science for All Americans*. New York: Oxford University Press. Available at http://www.project2061.org/publications/sfaa/online/sfaatoc.htm. Accessed March 8, 2013.

AAC&U (Association of American Colleges and Universities). (2007). *College Learning for the New Global Century*. Washington, DC: AAC&U. Available at http://www.aacu.org/leap/documents/GlobalCentury_final.pdf. Accessed March 8, 2013.

DeHaan, R.L. (2005). The Impending Revolution in Undergraduate Science Education. *Journal of Science Education and Technology*, 14 (2): 253–269.

Donald, J.G. (2002). *Learning to Think: Disciplinary Perspectives*. San Francisco: Jossey-Bass.

Eberlein, T., Kampmeier, J., Minderhout, V., Moog, R.S., Platt, T., Varma-Nelson, P., and White, H.B. (2008). Pedagogies of Engagement in Science. A Comparison of PBL, POGIL, and PLTL. *Biochemistry and Molecular Biology Education*, 36 (4): 262–273.

Gould, L.M. (1945). Science and the Other Humanities. Carleton College Presidential Inaugural Address, *Carleton College Bulletin*, 42 (2): np.

Hake, R.R. (2000). The General Population's Ignorance of Science Related Societal Issues: Challenges for the University. *AAPT Announcer*, 30 (2): 105.

Handelsman, J., Ebert-May, D., Beichner, R., Bruns, P., Chang, A., DeHaan, R., Gentile, J., Lauffer, S., Stewart, J., Tilghman, S., and Wood, W. (2004). Scientific Teaching. *Science*, April 23, pp. 521–522.

Huber, M.T,, Brown, C., Hutchings, P., Gale, R., Miller, R., and Breen, M. (eds.). (2007). *Integrative Learning: Opportunities to Connect*. Public Report of the Integrative Learning Project, Association of American Colleges and Universities and Carnegie Foundation for the Advancement of Teaching. Stanford, CA: Carnegie Foundation for the Advancement of Teaching. Available at http://www.carnegiefoundation.org/ilp/. Accessed March 8, 2013.

Hudes, I. and Moriber, G. (1971). A Liberal Arts Science Course for the 1970s. *Peabody Journal of Education*, 48 (2): 161–166.

LC National Center (Learning Communities National Resource Center). (2013). LC National Center. Available at http://www.evergreen.edu/washcenter/project.asp?pid=73. Accessed March 8, 2013.

Mestre, J. (2005). Facts and Myths about Pedagogies of Engagement in Science Learning. *Peer Review*, 7 (2): 24–27.

Michaelsen, L.K., Knight, A.B., and Fink, L.D. (2004). *Team-Based Learning. A Transformative Use of Small Groups in College Teaching*. Sterling, VA: Stylus.

Narum, J. (2008). Transforming Undergraduate Programs in Science, Technology, Engineering, and Math: Looking Back and Looking Ahead. *Liberal Education*, 94 (2): 12–17. Available at http://www.aacu.org/liberaleducation/le-sp08/le-sp08_Narum.cfm. Accessed March 8, 2013.

National Center for Case Study Teaching in Science (2013). National Center, SUNY Buffalo. Available at http://sciencecases.lib.buffalo.edu/cs/. Accessed March 8, 2013.

Nelson, C.E. (1989). Skewered on the Unicorn's Horn: The Illusion of Tragic Tradeoff between Content and Critical Thinking in the Teaching of Science. In L. Crow (ed.), *Enhancing Critical Thinking in the Sciences*, pp. 17–27. Washington, DC: Society for College Science Teachers.

NRC (National Research Council). (1996). *From Analysis to Action: Undergraduate Education in Science, Mathematics, Engineering, and Technology.* Report of a Convocation. Washington, DC: National Academies Press. Available at http://www.nap.edu/catalog.php?record_id=9128#toc. Accessed March 8, 2013.

NRC (National Research Council). (1999). *Transforming Undergraduate Education in Science, Mathematics, Engineering, and Technology.* Committee on Undergraduate Science Education, Center for Science, Mathematics, and Engineering Education. Washington, DC: National Academies Press. Available at http://www.nap.edu/catalog.php?record_id=6453#toc. Accessed March 8, 2013.

NSF (National Science Foundation). (1996). *Shaping the Future: New Expectations for Undergraduate Education in Science, Mathematics, Engineering, and Technology.* NSF No. 96–139. Washington, DC: National Science Foundation. Available at http://www.nsf.gov/pubs/1998/nsf98128/nsf98128.pdf. Accessed March 8, 2013.

PBL (Problem Based Learning). (2013). PBL, University of Delaware. Available at http://www.udel.edu/inst/. Accessed March 8, 2013.

PKAL (Project Kaleidoscope). (2002). *Report on Reports I.* Washington, DC: Project Kaleidoscope. Available at http://www.pkal.org/documents/ReportOnReports.cfm. Accessed March 8, 2013.

PKAL (Project Kaleidoscope). (2008). Pedagogies of Engagement Project. http://www.pkal.org/activities/PKALPhaseVI.cfm. Accessed March 8, 2013.

PLTL (Peer-Led Team Learning). (2013). PLTL. Available at http://www.pltl.org. Accessed March 8, 2013.

POGIL (Process Oriented Guided Inquiry Learning). (2013). POGIL. Available at http://www.pogil.org/. Accessed March 8, 2013.

Prince, M. and Felder, R. (2007). The Many Faces of Inductive Teaching and Learning. *Journal of College Science Teaching,* 36 (4): 14–20.

SERC (Science Education Research Center). (2013). SERC available at http://serc.carleton.edu/index.html. Accessed March 8, 2013.

Seymour, E. (2002). Tracking the Processes of Change in U.S. Undergraduate Education in Science, Mathematics, Engineering, and Technology. *Science Education,* 86 (1): 79–105.

Shulman, L.S. (2005). Pedagogies of Uncertainty. *Liberal Education,* 91 (2): 18–25. Available at http://www.aacu.org/liberaleducation/le-sp05/le-sp05feature2.cfm. Accessed March 8, 2013.

Sullivan, W.M. and Rosin, M.S. (2008). *A New Agenda for Higher Education: Shaping a Life of the Mind for Practice.* San Francisco: Jossey-Bass.

TBL (Team-Based Learning). (2013). TBL. Available at http://www.teambasedlearning.org. Accessed March 8, 2013.

Wieman, C. (2007). Why Not Try a Scientific Approach to Science Teaching? *Change,* 39 (5): 9–15.

2 From Student Learning to Teaching Foundations

Tricia A. Ferrett

CHAPTER 1 MADE a practical and moral case for connected science in higher education today, and it set this case within a historical context while looking forward to the kind of scholarship that will be required to understand the results of this educational philosophy when it is put into action. The heart of connected science lies in our work with students. Thus, I turn now to a discussion of how connected science learning can benefit them. Then, I sketch some emerging themes for teaching that are implied by this student learning discussion and the literature—and grow out of the case studies in this book.

The Value of Connected Science for Students

Today, the demands of a more global, connected, information-rich, and changing world places a higher premium on our students' abilities to integrate information and experience across traditional boundaries. Educators can no longer leave to chance the likelihood that students will make meaningful connections on their own across disciplines, courses, and experiences. College educators have often proclaimed that learning needs to be more than the sum of the parts. In fact, the liberal arts tradition intends for students to find high value through a broad education, connecting ideas and disciplinary fields in fruitful ways as they become lifelong learners. However, as faculty have come to better understand the challenges of making integrative and interdisciplinary connections in their own lives as teachers and scholars, they have become increasingly skeptical about whether college students can do this effectively enough on their own.

It is time to help students create more than the sum of the parts—with purpose and in structured community. The challenge now is to teach ourselves, through work with and study of our students, how to do this well. This book is our collective contribution to this effort. We describe here scholarly case studies of student learning in order to flesh out in a concrete way what connected science learning really looks like, how it works, the implications for teaching, and the inherent challenges and rewards for all involved.

Let's return to Jeff and Alice for a moment. How do they differ? Jeff was taking his learning to a new level in several ways. He was making connections between science concepts, decision making, and action. He clearly had become self-directed in his education. He was also developing an identity within an interdisciplinary learning community that was growing its capacity to act in the world with defined purpose. Jeff, more than Alice, is demonstrating characteristics of a connected science learner. Alice, while recognizing the importance of human issues for her research on the biochemistry of HIV-AIDS drugs, was constrained by the norms of a teaching culture that did not design for or encourage higher levels of connection making.

Clearly, the characteristics of the learner, the teacher, and the local culture are powerful factors in connected science learning and teaching. I tackle the first of these—the learner—by articulating student learning goals for connected science.

What are connected science learning goals for students? First, a disclaimer. This discussion is not meant to be complete. College educators are only beginning to articulate these kinds of learning goals for students, along with teaching practices that promote them. Thus, what follows is a synthetic and evidence-based attempt to articulate student learning goals for connected science. I draw freely on my own and others' ideas, across time, my experience, the literature, and a number of relevant reform projects. This original articulation will expand, interconnect, and deepen as educators and scholars think about, engage in, and study this kind of learning and teaching. Each instructor will emphasize a different subset of these learning goals, a variability reflected directly in the case studies in the volume.

I begin with some larger families of learning goals that apply inside and outside science learning, followed by narrowing to ones that are related to and grow from more science-rich contexts.

The Integrative Learner

Prior work on integrative learning is a fruitful place to start delineating connected science goals for students. The Carnegie Foundation for the Advancement of Teaching, in collaboration with AAC&U, hosted the Integrative Learning Project (ILP), engaging 10 institutions over three years. The final public ILP report (Huber et al., 2007) describes integrative learning as "developing the ability to make, recognize, and evaluate connections among disparate concepts, fields, or contexts" (np). This kind of learning has a creative dimension as well, inviting students to more actively engage themselves in defining their purpose and direction. The contexts and connections for integrative learning can vary enormously, as indicated by the stories of Jeff and Alice. The final ILP report is a rich source of information on how colleges and universities are attempting to develop the integrative capacities of their students over the college experience. The body of work in this book originates out of the 2005 Carnegie Scholar cohort, a group of 21 teaching scholars across disciplines investigating integrative learning in their own classrooms. Their project work, reported in online snapshots (Carnegie Scholars, 2006), complemented the Integrative Learning Project.

Huber and Hutchings (2004) write that integrative learners are intentional about all aspects of their learning. This mirrors the process of "intentional" teaching, where instructors develop learning environments that help students integrate and connect. An intentional learner, first of all, is purposeful, using a larger purpose to keep his learning on track. Connected science involves learning science in the context of a human purpose with enough open space so students can, to some extent, define their own purpose within boundaries set by both the context and the instructor. Related to this, an integrative learner is metacognitive (Huber and Hutchings, 2004)—she has developed skills to track her own thinking processes, using this self-awareness to monitor and refine her learning trajectory. A novice of metacognitive learning is getting practice in self-direction with regard to explicit learning goals, in learning when to ask for help, in monitoring and reflecting on her own efforts, and in making choices that promote learning. In fact, reflection is another major theme of integrative learning (Huber and Hutchings, 2004). Reflection puts "multiple perspectives into play with each other to produce insight" (Yancey, 1998, p. 6). This reflective process is central to integrative learning and connected science, where the burden for integrating ideas is frequently placed on the student.

There are an overwhelming number of ways to think about integrative learning—with multiple settings, pedagogies, and perspectives to enact or connect. For example, most of the authors in this volume were involved in a Carnegie Scholar community exercise to hash out a definition of integrative learning. We were asked to attach Post-its with brief phrases describing integrative learning to a whiteboard over a week. Once this became a collection of several hundred multicolored bits, we were asked to group them into broad categories. The final whiteboard displayed no less than 32 categories! The categories, for example, connected student learning to community service, other disciplines, formation of professional identity, forced-decision case studies, real-world problems, student experience and background, basic kinds of literacy, and so much more. This variety is also conveyed in the articles in the Summer–Fall 2005 issue of *Peer Review*, where the analysis, practice, and related research on integrative learning are addressed.

As I put forward a few ideas related to integrative learning that are most central in informing this volume of work, I acknowledge the depth of the concept and the way in which others will self-define their relationship to it. Likewise, the student learning goals for integrative learning are highly related, rich, and difficult to articulate. This begs for a scholarship of teaching and learning that addresses student learning using a range of methodologies, questions, theories, and perspectives. We hope to model in this book, at least in a beginning form, the kind of scholarship needed.

The Innovative and Flexible Thinker

Another body of work highly relevant to student learning goals for connected science comes from the literature on adaptive expertise. John Bransford and others distinguish routine expertise—where experts apply routine yet sophisticated processes to the same kinds of problems in a single domain or context—from adaptive expertise, where experts are able to adapt knowledge and skills to novel contexts while providing innovative solutions to problems (Bransford et al., 2000). Adaptive expertise involves, beyond the skills of routine expertise, flexible and creative approaches to problem solving across boundaries of traditional expertise. An adaptive expert is, in some sense, a fish out of water—a fish who relishes the challenge, thrill, and uncertainty of leaving its safe little fish tank for an uncertain but rich and creative adventure. Thus, the key abilities of adaptive expertise—innovation and flexibility of thought across boundaries—are highly resonant with those we seek to develop through connected science, especially in our nesting of student learning within a larger context that steps well outside a single scientific discipline. In addition, the intellectual process of integration itself is a creative one that, at its best, produces innovation of thought.

The Novice

Once college students start to develop into more integrative thinkers or adaptive experts, what will we find that they can do or understand better? This is an especially tricky question because connected science learning, like all learning, is a developmental process. So far little is understood, based on classroom evidence, about the intellectual development of integrative and innovative capacities in college. Thus, defining some key issues helps provide more shape to the above perspectives on learning goals for students who are *novices*—novices to science, the disciplines, and integrative thinking.

First, the attributes of an integrative learner or adaptive expert must be fostered from the first year of college (or earlier!) and not left to a capstone course or final-year project. Mary Taylor Huber and Pat Hutchings (2004) agree, warning that building capacities to integrate learning takes time and practice. Likewise, Bransford argues that adaptive expertise is not the same as routine expertise, and that some of the unique skills, attitudes, and habits of mind for adaptive expertise must be developed over time (Bransford, 2004).

With regard to intellectual development in the college years, a number of scholars have proposed schemas that indicate development happens roughly in stages, though not in a linear or even predictable way for a given student (Perry, 1970; Belenky, 1986;

King and Kitchener, 2004; Baxter Magolda, 2001). These schemas share in identifying common pivotal changes in development that relate fundamentally to a student's ability to argue from evidence and to recognize and cope with uncertainty. Nearly all the case studies in this book will describe learning that takes place in a zone of uncertainty. For example, the scheme by Patricia King and Karen Kitchener (2004) culminates in the learner taking a stand and making a personal commitment to an idea or action, despite significant uncertainty. This advanced stage of development requires a difficult move, the melding of knowledge and evidence-based reasoning with personal values. Some of the case studies that follow will stretch students to make a commitment, based on both science and values, and act in the changing world around them.

For connected science, this research on development helps us think about the implications of stretching our students, appropriately we hope, to cope with more complex systems and ill-defined problems. The possible solutions to these kinds of problems are many, providing an additional challenge, especially for first- and second-year college students. For example, for students in an early development stage who are still very much in the mode of thinking about ideas as right or wrong, addressing complicated problems in context may stretch them beyond their comfort zones of attitude or intellect. As educators, we need to remain keenly aware of this challenge as we design learning experiences for students. What does this mean? At first, we may need to reinforce the learning process more, guiding students so they can remain engaged and active without shutting down. As students get more comfortable and gain practice in the novice skills for connecting and integrating, the instructor's support may be pulled back a bit, guided by the instructor's wisdom and experience with his or her students.

Also critical to connected science are the developmental differences between novice and expert learners (Bransford et al., 2000). Part of the difficulty in articulating student learning outcomes for connected science is that as instructors, we naturally start from a position of expertise in our fields. The kinds of integrative capacities and products displayed by interdisciplinary experts—creation of new ideas and fields, application of a method in one field to another—are not likely to show up with novices. Instead, we need to watch our students for evidence that novices are beginning to make small yet productive moves—what I call *movelets*—in the direction of integrative or adaptive expertise. The case studies that follow provide fertile ground for identifying movelets in novices.

As with expert and novice skills, it is also helpful to make a distinction between interdisciplinary and integrative skills. Veronica Boix Mansilla's (2005) definition of interdisciplinary understanding creates a useful starting point, "interdisciplinary understanding [is] the capacity to integrate knowledge and modes of thinking in two or more *disciplines* or established areas of expertise to produce a cognitive advancement—such as explaining a phenomenon, solving a problem, or creating a product—in ways that would have been impossible or unlikely through single disciplinary means" (p. 16). Whereas the interdisciplinary expert will draw heavily on the discipline(s), evidence from the case studies that follow indicates that novices will draw on a wider range of perspectives—not all of them disciplinary. Given this tendency, it is particularly useful to rework Mansilla's definition into one for integrative understanding simply by replacing concepts connected to a discipline with ones related to the broader notion of a perspective: "*Integrative* understanding is the capacity to integrate knowledge and modes of thinking from two or more *perspectives* to produce a cognitive advancement—such as explaining a phenomenon, solving a problem, or creating a product—in ways that would have been impossible or unlikely through a single *perspective*." Where we have come to expect innovation from the integration of deeply understood disciplinary ideas or methods by experts, I have observed novice students simply noting two perspectives and discussing them and their connection, perhaps even on a superficial level. More sophisticated novices develop more purpose and some new ideas as they learn by connecting and integrating. No one has yet created and tested a "scale of sophistication" for integrative thinking appropriate for college students. This remains a challenge for us all. We can learn by documenting and naming the integrative movelets of our students. This growing sophistication may also be cast in terms of Benjamin Bloom's taxonomy (Bloom, 1956) for the development of intellectual skills. Bloom categorizes cognitive skills into groups that involve increasing levels of difficulty: knowledge, comprehension, application, analysis, synthesis, and evaluation. As we ask students to connect, integrate, and create, we are challenging them to operate toward the middle and higher end of Bloom's taxonomy. Again, developmental challenges arise, especially with novices.

This discussion about development really leads to the notion that the learning expectations for novices must be high, yet reachable and supported. Finding appropriate learning goals for novices, through the lens of development, stretches us outside our "expert" domain, beyond traditional notions of interdisciplinarity. Yet the case studies here will provide evidence that instructors can indeed help all college students develop their integrative capacities, and in multiple contexts. One major observation that emerges from our case studies is that our students, for their most fruitful integrative work, will likely draw on the course material and their own personal perspectives from prior knowledge, experience, and belief. At first, few if any of these personal perspectives will reflect a strong disciplinary grounding. As a student gains deeper experience in a discipline through a college major, more of the perspectives used in integrative learning become robustly disciplinary. At some point, and I will not propose when, enough disciplinary expertise develops that what I initially call integrative work morphs into interdisciplinary work. The key then is to invite younger students—novices—to draw on rich perspectives from the course and *themselves* so they can richly engage in connection-making that leads to cognitive advancements that enhance both their science learning and their integrative capacities.

The Connected Science Learner

All the dimensions of a learner discussed above—the integrative learner, the innovative and adaptive thinker, the novice—provide context for connected science learning goals. We gain further clarity on specific goals by considering how connected science relates to—and differs from—the ideals of two specific national science reform projects, particularly learning goals for connected science that relate to cognition, social skills, and attitude. This discussion intersects some aspects of classroom assessment in order to sharpen the vocabulary and concepts involved in articulating these goals.

I begin with a long-standing science reform project, Science Education for New Civic Engagements and Responsibilities (SENCER). SENCER (2008), which began in 2001 with NSF support, started with a focus on improving science education for undergraduates who were not science majors. As the project evolved, some instructors began to explore implications for the education of undergraduate science majors. Instructors created courses and curricula that connect science learning to critical civic issues. SENCER also helps instructors incorporate into course design results from research on human learning (Bransford et al., 2000) and principles of good assessment. Through their hosting of summer institutes, annual "reunion" symposia, and various other meetings, the project has built up a large and active reform community. The project has developed a wealth of online resources for course models, design templates, assessment, and civic issues (SENCER, 2012). SENCER has been around long enough that it has matured into a rich, connected, national reform community of practice.

SENCER is grounded in a well-articulated list of ideals that specify a vision for a learning environment where the student work is science rich, practical, integrative, and inquiry based. SENCER courses define a context around complex and contested public issues of consequence to create an open space for students to apply science, explore its limits, and deal with multiple perspectives. The context here is the driver for the learning, and the science content is drawn upon as needed. Thus, the ideals above imply a specific philosophy about the content and context of the student experience, along with some articulation of the role of the student as learner. Student learning goals focus on the application of knowledge and method, along with the powers, limits, and larger lessons of science. The learning environment itself asks the student to engage, discover, and apply. The role of the instructor as designer and facilitator of this rich learning environment is implied.

Connected science learning and teaching overlaps highly with the teach-through-public-issues approach promoted by the SENCER project. Both share a portrait of the student as active, engaged, and taking on a critical stance. Where connected science differs from SENCER is in the specificity with which we articulate student learning goals related to connection making and innovation, particularly through the process of integration of and reflection on multiple perspectives. As we discuss in these chapters, we also take a particular pedagogical point of view and attempt to articulate what we believe is much needed attention to the coherence between science content, context, and pedagogy. Another difference between SENCER and the connected science model is that SENCER is strongly oriented toward civic engagement, whereas several of the chapters in this volume reach beyond the realm of civic engagement to issues of professional education or Mansilla's notion of interdisciplinarity. Connected science learning includes but is not limited to public issues of civic engagement.

The second project of relevance is one in which I was heavily involved in the 1990s, the ChemConnections project (PKAL, 2004: ChemConnections 2013). This project grew out of two consortia, the ChemLinks Coalition (2008) and the ModularChem Consortium (2007), funded as partners in 1995 through the National Science Foundation's Systemic Change Initiative in Chemistry. The joint projects involved over 100 chemistry faculty from more than 42 universities, liberal arts colleges, and community colleges. Both consortia were interested in developing "modules" centered around real-world issues that required a chemistry perspective. Ultimately, a dozen modules were published (currently by Norton) on topics that included global warming, acid rain, automobile pollution, water treatment, and the origins of life. Each covered some "core content" in chemistry and took up several weeks of a college introductory chemistry course. In addition to connecting science learning to contemporary issues, the modules embedded active and collaborative learning activities in the student and instructor materials, making engaged pedagogies almost unavoidable. Finally, each module proceeded from a question, followed by smaller questions that composed a "story line" for inquiry. The modules allowed instructors to engage in guided inquiry, yet inquiry that at times could be turned over to students with small research activities.

One of the most important goals of the project was to "change ourselves" as educators. Much of what remains today from this work relates to the consequences of intense professional development that went on as faculty struggled to define, write, teach, revise, and assess student learning with modules. In addition, we tried hard to be clearer about student learning goals in general, including those for "thinking like a scientist." The whole premise, for example, of the ChemLinks project was to have students "do as chemists do" as they learned chemistry—actively, collaboratively, in a real-world context, and using the tools of scientific inquiry. Finally, the project included a large assessment and evaluation effort that drew some faculty into the world of assessment in a positive and generative way. Like SENCER, we thought hard about how to align instruction and learning activities with learning goals.

Looking back, one of the things we discovered in ChemConnections is that the modules that most appealed to faculty and students were the ones that grew from real-world questions that felt authentic, timely, and difficult to answer. This value we shared with SENCER and connected science. The best-selling module, titled "What Should We Do about Global Warming?," is even more relevant today than it was in the mid-1990s. We were also trying to be clearer about an educational philosophy that addressed the dissatisfaction of a generation of chemistry educators who were tired of teaching from textbooks that had changed little in decades. The contextual real-world questions—the module titles—drove the inquiry, active and collaborative activities (pedagogy) were an essential part of that inquiry, and the core chemistry content was chosen based on traditional teaching values and the issue at hand. Connected science also draws on this sense of authentic inquiry. However, during the project, we were not as explicit as we might have been about the notion of aligning content, context, and pedagogy. Some module authors discussed this alignment issue often but did not convey it to others in any significant way, Furthermore, teaching a module was difficult for many instructors to do, in both the short and long term. It required changes in multiple dimensions of teaching, and it went against the grain of department and professional tradition in many cases. In short, the project did not deal well enough with the issues of department and institutional culture, sustainability of the approach, and the challenges that arose from the magnitude of change required. There were also shortcomings in dissemination; the robust results on student learning remain unpublished today, leaving key evidence from the project mostly out of sight.

This brief discussion of the SENCER and ChemConnections projects shows both the lineage and a few challenges with regard to moving forward with connected science approaches. These two projects helped us think more clearly about cultural change, goals for student learning, "active learning" pedagogies, faculty development, and assessment. Both projects tied learning to real-world applications and attempted to engage students in more active inquiry. Truman Schwartz (2006) traces and analyzes these projects, along with their impressive ancestors—the high school *Chem-Com: Chemistry in the Community* curriculum and the American Chemical Society *Chemistry in Context* project for college students. These approaches, and many others, laid the essential groundwork for creating the territory for connected science teaching and learning.

What will students do and understand as they develop as connected science learners? How will instructors recognize their growth? One way to organize our thinking about outcomes is to first recognize that they fall into at least three major categories: attitudinal, cognitive, and social. On the attitudinal side, characteristics related to openness, playfulness, and risk taking begin to appear as students become more comfortable with boundary crossing and connection making. Students begin to ask questions that open the door to exploration of possible new connections. They may start to actively search for relevant approaches and perspectives beyond the ones at hand. They may start to self-define a purpose for themselves within the bounds of an integrative assignment or project. In the previous chapter, our friend Alice found purpose in her senior project by aiming to propose the structure for a more effective drug for AIDS. Students' readiness to adopt these attitudes likely depends on both their prior experience and personal inclinations and on their instructors' ability to create a learning environment that promotes, supports, and rewards such attitudes.

On the cognitive side, students may start to move from simple explanations to complex, multicausal ones, particularly when faced with a complex system or problem. They may start to make better use of tools for integrative thinking, some of which have been beautifully articulated by Jack Mino through a study of student writing and related "link alouds" in learning communities at his community college. Mino (2006), an author in this volume, finds that students use, often without conscious intention at first, "mechanisms of integration" such as personal experience, comparisons, posing integrative questions, metaphor and analogy, embedded quotes, and theory application across fields to gain new insight through a creative exploration of connections. My own experience indicates that students may first start to recognize and then later discuss the tensions between two perspectives on an issue. At higher levels of sophistication, students may actually engage in conceptual innovation, producing knowledge that is new to them (if not to the teacher or expert). Alice was able to integrate her knowledge of biochemistry to design a new molecular drug for treatment of HIV. Though the new drug remains untested, this is a nice example of knowledge production.

With regard to social interactions, communication and learning across differences in student backgrounds, majors, and belief systems are central to creating a rich learning environment for connected science. Recall Jeff's learning community in the prior chapter. On the first kayak trip, Jeff and his peers began to develop skills for listening, reflecting back, translating for others a concept in a field or discipline, and exploring questions as a group. These kinds of interactive skills are central to interdisciplinary research by experts (Mansilla, 2004). They are also quite difficult for experts to develop. What we are after here is an articulation of what novices in integrative learning do when they begin their journey in the social realm of learning. For example, seminars and learning communities, in particular, place great value on the development of these skills—and the ensuing benefits in learning. Ongoing work on social pedagogies is attempting to get at the essential aspects of this kind of teaching, the connection to adaptive expertise, and its benefits for students (Teagle Foundation, 2007).

Finally, given this rich set of learning goals, how will instructors know if students are reaching them? Research by Mansilla and her collaborators provides an entry into how to think about evaluation of student work that is integrative and interdisciplinary. Mansilla argues that interdisciplinary writing, for example, can be evaluated by instructors using a rubric that includes the following four criteria: a sense of purpose, the ability to integrate and produce a "cognitive advancement," disciplinary grounding, and a critical stance (Mansilla, 2005; Mansilla and Duraising, 2007). These evaluation criteria for student work link back nicely to our conceptual focus on purpose, integration, reflection, and intentional learning. For connected science and integrative learning, Mansilla's "disciplinary grounding" for novices in the sciences and disciplines can be reworked into "perspective grounding." In this case, instructors would want to look for appropriate integrity in the use of course material and other perspectives (some but not all of which will be disciplinary or even scientific) that a student brings to bear on her or his work. Ross Miller (2005) has summarized several approaches and assessment rubrics from this and other research, providing faculty with a place to start with assessment for integrative learning.

Teaching Themes and the Connected Science Learner

Related to the benefits for students highlighted above, some common themes emerge from our case studies related to connected science teaching. I briefly raise here four themes to consider through the following case studies. The case studies pay more or less attention to these issues, in varying ways, as the authors discuss results on connected science student learning.

First, instructors are paying more attention to building integrative habits of mind. Instructors will no doubt find their own ways of expressing the connected science learning goals noted aboveand other goals for their students' learning. As instructors explicitly design for integrative learning and connected science, they are adding new goals while discarding others. These choices, some of which are difficult, drive the design of integrative courses and assignments in very substantial and new ways. In a given course or assignment, not all of the goals are relevant. Instructors, as always, choose and adapt the ones of most importance to their students and contexts.

Second, instructors are paying more attention to research on human learning. This research has already provided the STEM community with a robust set of principles to guide course design. In addition to ideas about intellectual development noted above, the book *How People Learn* (Bransford et al., 2000) has been a focus of instructor development activities in STEM reform projects in the United States that include SENCER (2013) and Project Kaleidoscope (PKAL, 2006). This work, and research by others (e.g., Berge, 2002), indicates that effective learning is active, social, and reflective and occurs in a context. Learners build on preexisting knowledge and belief; teaching must raise and confront these aspects. Furthermore, the research on novices and experts indicates that teaching within an explicit conceptual framework helps learners better organize, retrieve, and apply their knowledge. Research on learning also highlights the important role of metacognition, as noted above. Connected science and many other STEM projects and approaches pay attention to this research. However, because instructors for connected science are explicitly asking their students to stretch, integrate, create, and connect, support and novice-expert distinctions are critically important to our success.

Third, connected science takes an original step forward—past existing reform movements and projects—by aspiring to align all major course aspects with a focus on real-world applications and the development of integrative capacities. Here is where I think connected science distinguishes itself the most from past efforts. First, this move toward coherence is a fairly radical stance to take in the recent history of undergraduate science education. Ideally, choices about course content, context, and pedagogy are made intentionally—and together. This is not easy. Yet some instructors, driven by the above benefits for their students, are beginning to select course context, content, and pedagogies that fit naturally with each other and their own goals for connected science learning. The course *context* for connected science relates to real-world import, providing both motivation and purpose to instructor and students. Ideally, the context is also naturally data rich, controversial, interdisciplinary, unfinished, roomy, and contemporary. This kind of context implies an open, student-influenced pedagogy, and a content that arises in natural and authentic ways given the contextual issue at hand.

The attributes of course content are more difficult to define and constrain. Many, though not all, the instructors in this volume select an object of study (e.g., an ecosystem, the human body) that is somewhat complex in nature, requiring students to integrate multiple layers of understanding into a fuller sense for how a system works. This leaning toward the study of complex systems and processes, in the context of a specific issue, creates especially fertile ground for students to explore science thinking, explanations, and systems that are integrated. Though instructors may well choose human or public issues that align with content they know must be part of a course, students then draw on the content on a need-to-know basis as they explore the issue at hand in a naturalistic way—much as a researcher does. Instructors are also asking their students to encounter a mix of existing and newly created knowledge, some of it generated by students through research or their own exploration of ideas.

The case studies will also bear out that the choice of pedagogy is of great consequence for effective connected science education. A later chapter in this volume by Richard Gale elaborates on what he calls "pedagogies of integration." Ideally, instructors design or adapt a pedagogical approach and related learning activities that are consistent with the kind of inquiry required for complex systems and contexts. Instructors can choose from a number of pedagogies that support students as they inquire, uncover, connect, create, stretch, dialogue, apply, deal with ambiguity and conflict, and sometimes make decisions of consequence and act in the community. Students will need convincing and support in order to thrive in a more open-ended learning setting where risk taking and intellectual play are emphasized. Work by José Feito (2007) notes the importance of creating a safe environment for "not knowing" that can lead to productive exploration of ideas, in his case through a seminar pedagogy. Some pedagogies can also allow for some self-definition of purpose on assignments and culminating projects if the learning is to be partly student driven. The pedagogical choices made by authors in this volume—including service learning, learning by teaching, learning communities, and seminar formats—also tend to favor learning that is social and highly relational, as in Jeff's case with his sustainability learning community. When learning goals focus on the exploration of multiple perspectives, peers can be a great source of ideas and feedback. Consistent with all this, an integrative pedagogy often takes a stand on the roles (some of them shared) of the teacher, learner, and peers. Finally, the notion that instructors can stand back and "gift the ownership of learning to the learner" (Malone, 2002, np) is a good fit for connected science learning.

Thus, what I am suggesting is a highly integrated and intentional approach to course design that attempts to optimize the alignment between content, context, and pedagogy. This requires instructors to take on a more "systems view" of their courses and the relationship between the essential course elements. One could also play the same alignment game with other central course elements, for example with the roles of the course learner, teacher, peers, and external consultants. As instructors gain experience with thinking more coherently about course content, context, and pedagogy, they will become more intentional teachers for integrative learning (Huber and Hutchings, 2004). This idea of integrated course design has been discussed by Elaine Seymour (2002) and Dee Fink (2003). Fink's book is a guide to the process of integrated course design for college instructors. We resonate with this kind of design process—but supplement it with a close, scholarly examination of what the courses and student learning actually look like. The case studies in this book also attempt to focus design issues on the bottlenecks and challenges that emerge with real student learning.

Finally, the connected science approach is committed to helping students develop and mature as whole individuals. This theme points back to ideas noted above on citizenship and the development of our students' self-purpose and integrative and reflective capacities for use on complex decisions in the real world. Indeed, we can find the heart of connected science teaching in Parker Palmer's words (1997, p. 11): "Good teachers possess a capacity for connectedness. They are able to weave a complex web of connections among themselves, their subjects, and their students so that students can learn to weave a world for themselves."

References

Baxter Magolda, M.B. (2001). *Making Their Own Way: Narratives for Transforming Higher Education to Promote Self-Development.* Sterling, VA: Stylus.

Belenky, M.F. (1986). *Women's Ways of Knowing: The Development of Self, Voice, and Mind.* New York, NY: Basic Books.

Berge, Z.L. (2002). Active, Interactive and Reflective Learning. *Quarterly Review of Distance Education,* 3 (2): 181–190.

Bloom, B.S. (1956). *Taxonomy of Educational Objectives, Handbook I: The Cognitive Domain.* New York: David McKay.

Bransford, J. (2004). *Thoughts on Adaptive Expertise.* Unpublished manuscript.

Bransford, J.D., Brown, A.L., and Cocking, R.R. (2000). *How People Learn: Brain, Mind, Experience, and School.* Expanded ed. Washington, DC: National Academy Press.

Carnegie Scholars. (2006). Final Project Snapshots for 2005 Cohort. Carnegie Foundation. Available at http://gallery.carnegiefoundation.org/gallery_of_tl/castl_he.html. Accessed March 8, 2013.

ChemConnections. (2013). W.W. Norton. Available at http://books.wwnorton.com/books/search/aspx?searchtext=ChemConnections. Accessed March 8, 2013.

ChemLinks Coalition (2008). ChemLinks Project. Available at http://chemlinks.beloit.edu. Accessed March 8, 2013.

Feito, J.A. (2007). Allowing Not-Knowing in a Dialogic Discussion. *International Journal for the Scholarship of Teaching and Learning*, 1 (1). Available at http://academics.georgia southern.edu/ijsotl/2007_v1n1.htm. Accessed March 8, 2013.

Fink, D. (2003). *Creating Significant Learning Experiences: An Integrated Approach to Designing College Courses*. San Francisco: Jossey-Bass.

Huber, M.T. and Hutchings, P. (2004). *Integrative Learning: Mapping the Terrain*. Washington, DC: Association of American Colleges and Universities.

Huber, M.T., Brown, C., Hutchings, P., Gale, R., Miller, R., and Breen, M., eds. (2007). *Integrative Learning: Opportunities to Connect*. Public Report of the Integrative Learning Project, Association of American Colleges and Universities and Carnegie Foundation for the Advancement of Teaching. Stanford, CA: Carnegie Foundation for the Advancement of Teaching.

King, P.M. and Kitchener, K.S. (2004). Reflective Judgment: Theory and Research on the Development of Epistemic Assumptions through Adulthood. *Educational Psychologist*, 39 (1): 5–18.

Malone, L. (2002). *Peer Critical Learning*. Carnegie Scholar Final Project Snapshot. Available at http://www.cfkeep.org/html/snapshot.php?id=27974468309408. Accessed March 8, 2013.

Mansilla, V.B. (2004). Interdisciplinary Work at the Frontier, an Empirical Examination of Expert Interdisciplinary Epistemologies. *Issues in Interdisciplinary Studies*, 24: 1–31.

Mansilla, V.B. (2005). Assessing Student Work at Disciplinary Crossroads. *Change*, 37 (January–February): 14–21.

Mansilla, V.B. and Duraising, E.D. (2007). Targeted Assessment of Students' Interdisciplinary Work: Empirically Grounded Framework Proposed. *Journal of Higher Education*, 7 (2): 215–237.

Miller, R. (2005). Integrative Learning and Assessment. *Peer Review*, 7 (4): 11–14.

Mino, J. (2006). *The Link Aloud: Making Interdisciplinary Learning Visible and Audible*. 2005 Carnegie Scholar Final Project Snapshot. Available at http://gallery.carnegiefoundation .org/gallery_of_tl/making_interdisciplinary_connections.html. Accessed March 8, 2013.

Modular Chemistry Consortium (2007). MC^2 Project. Available at http://mc2.cchem.berkeley. edu/. Accessed March 8, 2013.

Palmer, P.J. (1997). *The Courage to Teach: Exploring the Inner Landscape of a Teacher's Life*. San Francisco: Jossey-Bass.

Perry, W.G., Jr. (1970). *Forms of Intellectual and Ethical Development in the College Years: A Scheme*. New York: Holt, Reinhart, and Winston.

PKAL (Project Kaleidoscope). (2004). *Modules as a Tool, 21st Century Pedagogies*. Vol. 4, *What Works, What Matters, What Lasts*. Washington, DC: Project Kaleidoscope. Available at http://pkal.org/documents/Vol4ModulesAsAToolPedagogies.cfm. Accessed March 8, 2013.

PKAL (Project Kaleidoscope). (2006). Translating *How People Learn* into a Roadmap for Institutional Transformation. PKAL. Available at http://www.pkal.org/collections/HPL.cfm. Accessed March 8, 2013.

Schwartz, A.T. (2006). Contextualized Chemistry Education: The American Experience. *International Journal of Science Education*, 28 (9): 977–998.

SENCER (Science Education for New Civic Engagements and Responsibilities). (2013). SENCER Project. Available at http://www.sencer.net/. Accessed March 8, 2013.

Seymour, E. (2002). Tracking the Processes of Change in U.S. Undergraduate Education in Science, Mathematics, Engineering, and Technology. *Science Education*, 86 (1): 79–105.

Teagle Foundation (2007). *Defining Social Pedagogies and Their Relevance to Liberal Learning.* Teagle's Fresh Thinking Initiative. Available at http://www.teaglefoundation.org/Grant making/Grantees/default/?gg=796&rfp=387&o=0. Accessed March 8, 2013.

Yancey, K.B. (1998). *Reflection in the Writing Classroom*. Logan, UT: Utah State University Press.

PART II

COURSES THAT FOSTER INTEGRATIVE LEARNING

3 Public Health and Biochemistry

Connecting Content, Issues, and Values for Majors

Matthew A. Fisher

ONE OF THE challenges when incorporating integrative learning experiences in the undergraduate science curriculum for majors is the widely held perception by faculty that such changes would require significant sacrifices in the content that students learn. In my experience, however, changes made in a biochemistry course sequence for biochemistry, biology, and chemistry majors allowed the introduction of integrative learning opportunities without the loss of course disciplinary content. The revised sequence accomplished this goal by framing course content in the context of pressing public health issues such as Alzheimer's disease, HIV/AIDS, and influenza. The revised courses challenged students to look at these issues from the perspective of biochemistry as well as other disciplines, their personal values, and institutional values.

Students have no problem anticipating that biochemistry will have a significant connection to what they are interested in, care about, and encounter on a daily basis; however, biochemistry textbooks and courses have traditionally steered clear of non-disciplinary discussions of the complexity of diseases such as AIDS and malaria or of malnutrition. The content of undergraduate biochemistry courses is thus most commonly presented in a manner that is largely disconnected from real-world contexts. Without a textbook or course pedagogy that makes clear these connections and establishes a context for knowledge, the stage is set for a pathology of learning that Lee Shulman (1999) has described as inertia—an inability to use what has been learned. There are several studies that clearly and persuasively argue that traditional curricula in chemistry (see Cooper, 2010 for a summary) and biology (National Research Council, 2003) include too much content and the result is often the inertia that Shulman describes. Upper-level undergraduate biochemistry courses are typically very content intensive; my fall course looks at protein structure and function, enzyme function (kinetics, mechanism, regulation), and roughly a half dozen distinct metabolic pathways (reactions involved, overall energetics, regulation) related to how the human body metabolizes carbohydrates, fats, and proteins. The spring course examines some different aspects of protein structure, glycoproteins, membrane structure and dynamics, transport, signal transduction, nucleic acid structure, and several processes central to nucleic acid biochemistry (replication, DNA repair, transcription, translation).

Shulman (1997) points out that the solution to this pathology of inertia is to ensure that "the subject-matter content to be learned is *generative*, essential and pivotal to the discipline or interdiscipline under study, and can yield new understandings and/or serve as the basis for future learning of content, processes, and dispositions" (p. 493). How might we ensure that the content learned by science majors in undergraduate biochemistry is "generative"?

Considering more explicitly the affective dimensions of learning is likely to be an important part of answering this question. Elizabeth Barkley (2010) in *Student Engagement Techniques* provides an exceptional summary of how we can create synergy in learning environments by engaging students not only cognitively but also through affect and body (psychomotor) dimensions of learning. More extended discussions of the important role that the affective domain plays in learning, memory, and cognitive function can be found in the National Teaching and Learning Forum (Nuhfer, 2008a, 2008b; Rhem, 2008a, 2008b) and works by Damasio (1994, 1999).

Within the context of the broader learning goals for higher education, such as those put forth by AAC&U's *College Learning for the New Global Century* (2007) and Derek Bok's *Our Underachieving Colleges* (2005), generative content becomes even more important. AAC&U has established the following essential learning outcomes: knowledge of human cultures and the physical and natural world, intellectual and practical skills, personal and social responsibility, and integrative learning. These goals echo what is increasingly heard within the scientific community itself. In his 2002 *Science* article "World Poverty and Hunger—the Challenge for Science," Ismail Serageldin provided a charge to the scientific community to educate scientists in ways that make visible the connections of basic science to the issues and concerns of a global society.

> For science to realize its full promise and become the primary force for change in the world, it requires that scientists work to 1) engage scientific research in the pressing issues of our time; 2) abolish hunger and reduce poverty; 3) promote a scientific outlook and the values of science; 4) build real partnerships with the scientists in the South. . . . All of that, however, requires our joint commitment as scientists to work for the benefit of the entire human family, not just the privileged minority who are lucky enough to live in the most advanced industrial societies. . . . But scientists' voices must be heard loudly and clearly in the national discourse of their

own societies. This absence not only severs science from its salutary effect on the modernization of societies, but also undermines the public support necessary for its pursuit. (p. 57)

To prepare undergraduate science majors for the responsibilities that Serageldin describes requires that students make connections between scientific concepts studied in courses for their major, larger social issues, and their own personal values and willingness to engage in moral and civic issues. I have written elsewhere at greater length (Fisher, 2010) about the challenges of educating for citizenship within the framework of an undergraduate biochemistry course. Clearly integrative learning will require that courses for science majors present content in a way that is generative.

The connected science conceptual breakthrough for me was realizing that connecting biochemistry content to context didn't require "covering" additional material but rather recasting the foundational concepts in a different way that was also generative. Consider the topic of globular protein structure. Most textbooks present the basic concepts of protein structure—the packing of secondary structure, the stability of the final fold, the prevalence of common structural motifs—using proteins such as myoglobin (oxygen storage), hemoglobin (oxygen transport), cytochrome c (electron transport), or some enzymes from central metabolic pathways such as glycolysis. Yet many of the same concepts can be illustrated with two proteins—hemagglutinin and neuraminidase—from the influenza virus. The difference is that using the viral proteins as illustrative examples to "teach through" to the underlying biochemical concepts provides a clear connection to influenza epidemics and the ongoing concern about bird flu as a context for the content. In that context, it is a small and natural step to includes readings such as Michael Specter's "Nature's Bioterrorist" from *The New Yorker* and Malcolm Gladwell's "The Dead Zone" and "The Scourge You Know," also from *The New Yorker.* Now, with influenza as the context, a number of other questions and issues and disciplines become relevant to a biochemistry class.

The approach that I have just described is, in fact, not new. It lies at the heart of the SENCER Project (2013), described in chapter 2. SENCER's approach is teaching "to" basic, canonical science and mathematics "through" complex, capacious, often unsolved issues of civic consequence. While much of SENCER's work in the project's first decade has focused on courses for nonscience majors, more and more conversations at the annual SENCER Summer Institute have revolved around the potential benefits of using this approach in courses for science majors. Having been involved with SENCER since 2002 and having used this approach to teach several courses for nonscience majors, I thought more about employing components of the SENCER model in my biochemistry classes. I also realized that many of the illustrative examples textbooks use for important biochemical concepts, examples that are very much disconnected from real world issues and thus fail to function as "generative content," could be replaced with examples drawn from public health topics. So in the fall of 2005 I redesigned my courses in three ways:

1. 1] Wherever possible, illustrative examples used in class would be drawn from public health topics rather than the typical examples found in textbooks that had been used for many years.
2. 2] At various points in the course, I would ask students to read and respond to articles that focused on the broader societal context of these public health issues.
3. 3] Students would work in small groups to develop a presentation on the biochemistry of a public health topic of their own choosing.

My goal was to provide openings or opportunities in the course for students to make connections among the information, ideas, and perspectives in this course and those in other courses or areas outside the natural sciences. The idea for the public health presentation was inspired by comments from other members of my cohort of Carnegie Scholars, who posed two challenges to me: find a way to give students some authority to make choices and explore what interests them in the course and provide students with an opportunity to demonstrate the connections they make.

Table 3.1 lists the various public health topics that I have incorporated into both semesters of biochemistry. Introductory readings, such as the Gladwell (1997, 2001) and Specter (2005) articles, provide the broader context for the topic and are connected to an assignment that involves some reflective writing. For example, in the readings for influenza, the reflective writing assignment asks students to identify the particular points they see as important to what the author is trying to communicate in each article, identify what questions they have about each reading, and reflect on the ideas in the reading as they relate to the core values central to both the Benedictine monastic tradition that shapes Saint Vincent College and the college's own understanding of its mission, articulated as community, care, hospitality, and stewardship. The same topics serve as the source of illustrative examples for basic concepts in the course; for instance, the two influenza viral proteins serve as examples when we introduce and examine more closely basic concepts in protein structure and stability. Finally, the take-home question for the exam at the end of each section is almost always related to the topic we have been using as the context. Whenever possible, the take-home question contains sections that focus on the broader issue and relevant perspectives as well as sections that focus on the basic biochemistry.

The content that students encounter in the public health version of the upper-level biochemistry sequence at Saint Vincent College includes the same topics as most undergraduate biochemistry courses. In fact, the only topic that I had to remove to "make space" was photosynthesis.

As noted earlier, faculty who teach courses for science majors are concerned about potential losses of content with any change in pedagogy. This concern is not unfounded, as recent documents such as the report *Scientific Foundations for Future Physicians* (HHMI-AAMC, 2009) make clear. This report lays out recommendations for preparing undergraduates to continue on to medical school that include 8 broad competencies and 37 major learning objectives. Similar expectations are held

Table 3.1. Cases and Themes Used in Both Semesters of Biochemistry

Public Health Issue	Semester	Basic Concepts Covered
Alzheimer's disease	Fall	Protein structure, stability, and folding illustrated by amyloid β protein
HIV/AIDS	Fall	Enzyme kinetics and mechanism illustrated by HIV proteaese
Diabetes and malnutrition	Fall	Central metabolic pathways (glycolysis, citric acid cycle, ATP synthesis), gluconeogenesis, glycogen metabolism, fat metabolism, and integration of metabolic pathways by looking at the biochemical basis and consequences of diabetes, obesity, and malnutrition
Avian influenza	Spring	Protein structure, stability, function, and glycoproteins by examining the functions of hemagglutinin and neuraminidase from influenza virus, designing a flu vaccine, and designing new antivirals
Multidrug-resistant tuberculosis	Spring	Transport across membranes illustrated by the molecular mechanisms for multidrug resistance in tuberculosis
Neuropsychiatric conditions	Spring	Signal transduction by various pathways illustrated by the biochemical effects of neurotransmitters and neuropsychiatric drugs
Cancer and the environment	Spring	Signal transduction, DNA replication, and transcription (both mechanism and regulation) illustrated through the biochemical consequences of damaging DNA or incorrectly turning on/off proteins (such as estrogen receptor) involved in signal transduction or regulation of transcription

by graduate and other professional programs such as dental and veterinary medicine schools. Demonstrating that a change in pedagogy will not result in significant loss of content is an important task for any reform effort.

Do students develop the same level of understanding when we teach biochemistry in a more "connected" manner using public health topics? The final exam in each course is cumulative, and has been essentially the same exam each year except for small revisions to reflect how each offering of the course actually unfolds. Questions on the exam ask students to define some concepts and terms, explain interrelationships between concepts, apply concepts as part of analyzing data sets, and compare related concepts. From the fall of 2000 to the spring of 2005, the average on the final exam was 80 ± 14 in the fall course (39 students total) and 81 ± 11 in the spring course (34 students). In the three years beginning with the fall of 2005 that I have taught the biochemistry sequence with a public health context, the final exam averages were 81 ± 8 in the fall course (44 students total) and 79 ± 10 in the spring course (35 students total). These comparative final exam scores offer critical evidence to counter the misconception that biochemistry courses structured to support integrative learning compromise disciplinary content coverage and decrease student learning of biochemistry.

The connection between content (biochemical concepts) and context (public health) now integral to the course has resulted in the same level of content understanding by students and has elicited a positive student response that transcends the course. In the spring of 2005 I shared with one of my former students, a biochemistry major now attending medical school at the University of Michigan, how I planned to change the course. She responded in an e-mail:

> I am even more convinced that this is something which would truly have a positive impact on undergraduate curriculum, and not just for students who intend to enter a medical field. I think that more and more, we need to be able to relate what we learn in the classroom to the outside world. This makes the classroom lessons more tangible, and more applicable, and honestly, I think it makes the learners more interested. Some of my favorite questions from your class were the ones where you presented us with a "real life" scenario, and told us to describe the underlying biochemical change. To this day, I remember those processes and the logic behind them, but the Krebs cycle that I merely memorized is but a distant memory.

Below are some student responses to the question "At what moment in class during the past (two to four) weeks did you feel most engaged with what was happening?"

> I believe that the public health aspect does a very good job of this. This deepens my understanding in terms of application while at the same time serves to increase interest by showing the real world application. (Spring 2007)

> I really like the cases and applications because it takes you a step outside and this way we pool stuff together to make sure we can know what's going on. (Fall 2007)

> I feel most engaged when we discuss topics such as health issues—AIDS. Fall 2007

I have had similar conversations with numerous students over the past three years. The integration of context and content in such a way that the science connects to broader issues has helped foster a learning environment in my classes where students are much more engaged affectively at the same time that they are engaged cognitively. The connection between affective engagement and motivation is quite clear; the National Research Coucil report *How People Learn* (2000) asserts that "[l]earners of all ages are more motivated when they can see the usefulness of what they are learning and when they can use that information to do something that has an impact on others—especially their local community" (p. 61). And in *The Art of Changing the Brain* (2002), James Zull summarizes a large body of biological and cognitive science research, noting that "feelings always affect reasoning and memory. This influence of feelings runs the entire gamut of possibilities. Feeling[s] can help us remember and make us forget. They can help us recall important events that did happen, but they can also trigger false memories. They are essential for reasoning, and they can hinder reasoning" (p. 86).

The public health projects that students completed on topics of their own choosing provide particularly rich examples of the connections that science majors can make. The project involves presenting an overview of the biochemistry central to a public health topic chosen by each small group of students as well as perspectives outside biochemistry that the group identifies as important to understanding the particular issue. The students submitted final projects electronically using the KEEP Toolkit, an open-source web-based tool developed by the Knowledge Media Lab at the Carnegie Foundation for the Advancement of Teaching (Iiyoshi and Richardson, 2008).[1] This means of presentation allows for consideration of broader audience, makes space in the course for creativity and utilization of web-based presentation skills, and provides evidence of students' integrative understanding.

The assignment guidelines stressed four essential elements:

1. Identifying and presenting an overview of some biochemistry central to the topic the group has chosen; since for most topics it is impossible for students to present an overview of all the biochemistry involved I encourage students to pick one or two aspects and focus on those.
2. Identifying one or two other perspectives outside of biochemistry and character- istics from those perspectives that are of notable significance in understanding the issue selected. Groups are expected to use one or more of the core values— community, care, hospitality, stewardship—viewed by our faculty as central to the Benedictine tradition as a 'critical lens' through which students approach these other perspectives.
3. Presenting what the group sees as important next steps to be taken in addressing the issue.
4. Incorporating the personal perspective/thoughts of each group member about this issue into the presentation.

The student projects provided a wide array of integrative understanding.[2] For example, one project investigating malaria demonstrated connections to a microbiology course and to several different public health agencies such as the World Health Organization and the Centers for Disease Control and Prevention. At the same time, this presenta- tion was weakened by the minimal incorporation of biochemistry and connections to disciplines outside science. A public health project focusing on Shigella showed a rich integration of biochemistry and personal perspectives, yet it did not develop the connections to other disciplines outside science as fully. A rabies public health project included some biochemistry and a rich incorporation of microbiology, ethics, and the personal interest of this particular group in veterinary medicine. Projects on river blindness and alcoholism showed a deep and connected use of biochemistry and other scientific perspectives (e.g., microbiology and physiology) as well as the disciplines of economics, sociology, and theology. In these public health projects, students revealed the connections they saw between their academic experience and general knowledge, personal perspectives, values and ethics, and the larger communities that influence the problems and solutions of humanity. And while the number, type, and depth of the connections varied, the biochemistry content knowledge was not significantly different from previous semesters, as evidenced by final exam scores, on tests that did not ask students to frame their understanding of basic biochemistry within the context of global public health issues.

The changes to the biochemistry course sequence described in this chapter required minimal sacrifice in the course content, and evidence showed that students were much more engaged affectively with the course material. As a result, students indicated that they had a greater sense of purpose in their learning and were beginning to integrate "fragments of knowledge" from courses in both the sciences and general education.

William Sullivan (2004) describes a model for professional education (such as in STEM fields) that builds on three "apprenticeships." The first is a cognitive apprenticeship where the student learns to think like a member of the profession. The second is a skill apprenticeship where the student practices the skills routinely used by members of the profession. The third is a moral apprenticeship that "teaches the skills and traits, along with the ethical comportment, social roles, and responsibilities, that mark the professional" (p.208). Revising these courses to provide more integrative learning opportunities allowed students to be more fully engaged in all the dimensions of professional education that Sullivan describes.

Notes

1. The KEEP Toolkit can now be found through MERLOT http://about.merlot.org/KEEP.html.
2. A sample of the work done by students from fall 2005 through spring 2008 can be found at http://contentbuilder.merlot.org/toolkit/users/mattfisher/publichealthprojects.

References

AAC&U (Association of American Colleges and Universities). (2007). *College Learning for the New Global Century*. Washington, DC: AAC&U.

Barkley, E. (2010). *Student Engagement Techniques: A Handbook for College Faculty*. San Francisco: Jossey-Bass.

Bok, D. (2005). *Our Underachieving Colleges: A Candid Look at How Much Students Learn and Why They Should Be Learning More*. Princeton, NJ: Princeton University Press.

Cooper, M. (2010). The Case for Reform of the Undergraduate General Chemistry Curriculum. *Journal of Chemical Education*, 87 (3): 231–223.

Damasio, A. (1994). *Descartes' Error: Emotion, Reason, and the Human Brain*. New York: Penguin.

Damasio, A. (1999). *The Feeling of What Happens: Body and Emotion in the Making of Consciousness*. New York: Harcourt.

Fisher, M.A. (2010). Educating for Scientific Knowledge, Awakening to a Citizen's Responsibility. In M.B. Smith, R.S. Nowacek, and J.L. Bernstein (eds.), *Citizenship across the Curriculum*, pp. 110–131. Bloomington: Indiana University Press.

Gladwell, M. (1997). The Dead Zone. *New Yorker*, September 29, pp. 52–64.

Gladwell, M. (2001). The Scourge You Know. *New Yorker*, October 29, pp. 34–35.

HHMI (Howard Hughes Medical Institute) and AAMC (Association of American Medical Colleges). (2009). *Scientific Foundations for Future Physicians: Report of the AAMC-HHMI Committee*. Washington, DC: Association of American Medical Colleges.

Iiyoshi, T. and Richardson, C.R. (2008). Promoting Technology-Enabled Knowledge Building and Sharing for Sustainable Open Educational Innovations. In T. Iiyoshi and M.S.V. Kumar (eds.), *Opening Up Education: The Collective Advancement of Education through Open Technology, Open Content, and Open Knowledge*, pp. 337–355. Cambridge, MA: Massachusetts Institute of Technology Press.

National Research Council. (2000). *How People Learn: Brain, Mind, Experience, and School*. Expanded ed. Washington, DC: National Academies Press.

National Research Council. (2003). *Bio 2010: Transforming Undergraduate Education for Future Research Biologists*. Washington, DC: National Academies Press.

Nuhfer, E. (2008a). The Affective Domain and the Formation of the Generator: Educating in Fractal Patterns XXIII. *National Teaching and Learning Forum*, 17 (2): 8–11.

Nuhfer, E. (2008b). The Affective Domain and the Formation of the Generator Part 2: Fractal Perspectives on the Importance of the Affective Domain: Educating in Fractal Patterns XXIV. *National Teaching and Learning Forum*, 17 (3): 9–11.

Rhem, J. (2008a). The Affect Issue. *National Teaching and Learning Forum*, 17 (2): 1–3.

Rhem, J. (2008b). The Affective Field. *National Teaching and Learning Forum*, 17 (2): 4–5.

SENCER (Science Education for New Civic Engagements and Responsibilities). (2013). http://www.sencer.net/. Accessed March 8, 2013.

Serageldin, I. (2002). World Poverty and Hunger—the Challenge for Science. *Science*, 296: 54–58.

Shulman, L.S. (1997). Communities of Learners and Teachers. In S. Wilson (ed.), *The Wisdom of Practice: Essays on Teaching, Learning, and Learning to Teach*, pp. 485–500. San Francisco: Jossey-Bass, 2004.

Shulman, L.S. (1999). Taking Learning Seriously. *Change*, 31 (4): 10–17.

Specter, M. (2005). Nature's Bioterrorist. *New Yorker*, February 28, pp. 50–61.

Sullivan, W.M. (2004). *Work and Integrity: The Crisis and Promise of Professionalism in America*. 2nd ed. San Francisco: Jossey-Bass.

Zull, J. (2002). *The Art of Changing the Brain: Enriching the Practice of Teaching by Exploring the Biology of Learning*. Sterling, VA: Stylus.

4 Designing to Make a Difference

Authentic Integration of Professional Skills in an Engineering Capstone Design Course

Gregory Kremer

Engineering Education and Capstone Design

What do engineers do, and how do they contribute to society? If you are unsure of your answer, you would be in the same position as the public in general and many potential and current engineering students in particular. In response to this situation, the National Academy of Engineering completed a study in 2008 titled *Changing the Conversation*. The study (NAE, 2008) found the most effective messages for helping a variety of audiences understand the role, importance, and career potential of engineering are:

1. Engineers make a world of difference.
2. Engineers are creative problem solvers.
3. Engineers help shape the future.
4. Engineering is essential to our health, happiness, and safety.

These messages communicate the transcendent promise and purpose of engineering—positively affecting the world by implementing new ideas and solutions. But does our current system of engineering education develop engineers with this mind-set and the abilities to make this vision a reality?

One valuable perspective on this question is presented in *Educating Engineers: Designing for the Future of the Field* (Shepherd et al., 2009), the third volume in the Preparation for the Professions Series by the Carnegie Foundation for the Advancement of Teaching. The book's authors assert that "today's engineering education is strong on imparting some kinds of knowledge but not very effective in preparing students to integrate their knowledge, skills, and identity as developing professionals" (p. 169). In other words, students know a lot about *how to do* engineering, but not enough about *how to be* an engineer. The downloadable summary of the book states it this way: "[T]he central lesson that emerged from the study is the imperative of teaching for professional practice—with *practice* understood as the complex, creative, responsible, contextually grounded activities that define the work of engineers at its best; and *professional* understood to describe those who can be entrusted with responsible judgment in the application of their expertise for the good of those they serve" (p. 7). The report's four principles for improving engineering education have a strong emphasis on integrative education and making connections, and in the downloadable summary the authors add "that social and ethical connections are as important, if not more so, as electrical and mechanical ones" (p. 9):

1. Provide a professional spine.
2. Teach key concepts for use and connection.
3. Integrate identity, knowledge, and skills through approximations to practice.
4. Place engineering in the world: encourage students to draw connections.

Many engineering educators have been working to contextualize engineering education and provide the type of empowering education that our students deserve and our society needs. The service learning community (www.ijsle.org) is one group of educators working to integrate the theory and practice of engineering using a reflective approach meant to provide graduates with a sense of purpose, a set of skills, and a way of thinking that are understood in connection to their lives and to the practice of engineering. On a broader level, accreditation guidelines have changed in the last decade to require programs to help students develop a variety of skills and perspectives important for professional practice, including an ability to work on multidisciplinary teams. The accreditation criteria also include this statement: "Students must be prepared for engineering practice through a curriculum culminating in a major design experience based on the knowledge and skills acquired in earlier course work" (ABET, 2011, p. 6). Another accreditation requirement is that programs must have an advisory board so that a broad perspective on professional practice is reflected in the program's educational objectives. One of the four objectives put forth by the Industrial Advisory Board for the Ohio University mechanical engineering program is to graduate mechanical engineers with skills to perform in the work environment, a task that includes such things as formal and informal communication, teamwork, project management, appreciation of engineering integration with business, ethical and effective decision making, and the ability for self-evaluation, leading to improvement. These skills for professional practice are often called professional skills, but recent studies have shown that while faculty understand the importance of professional skills and interpersonal skills in "how to be" an engineer, very few report explicitly teaching them (Matusovich et al., 2009). So it should not be surprising that a recent Academic Pathways study that took a longitudinal look at a range of engineering seniors found that 40 percent did not think their school experiences contributed significantly to their knowledge of engineering practice. These seniors had low confidence in their interpersonal skills and a mistaken impression that those skills are less important professionally, revealing a troubling disconnect between their college experience and the reality of engineering practice (Lord, 2010).

Team-based capstone design experiences are the normal response to the accreditation requirements and are a common place for developing and assessing the outcomes related to the skills and perspectives important for professional practice. *Capstone* implies a crowning achievement, and most capstone design teams are challenged to solve an open-ended problem that is of greater scope and complexity than any they have encountered in their prior academic preparation and that requires application and extension of previously learned knowledge and skills. A capstone design course is project centered, not topic centered, so lectures are infrequent and are driven by the skills and knowledge needed for the project. The teacher normally takes on a coaching or mentoring role, often interacting with individual teams in project meetings to address issues specific to their project (Shepherd et al., 2009). Some capstone projects are only "paper" designs that are not actually built and tested, but the most meaningful experiences come when students experience the entire design process, from problem definition through building and testing a device that solves the problem and providing technical support to the customer using their capstone project. The team aspect is an important opportunity to experience the importance of interpersonal communication, and most capstone design projects include a requirement for design reports and formal presentations. Most capstone design experiences have a dominant emphasis on bringing diverse technical skills together in a meaningful application, in part because of the strong foundation laid by the many courses in the curriculum that develop technical skills for how to do engineering. But few if any engineering programs have a similar foundation for the professional skills and perspectives important for engineering practice.

The term *connected capstone* will be used in this chapter to indicate a community-partnership capstone experience where intentional connections are made to integrate doing engineering with being an engineer. The pedagogical intent is to increase student appreciation of the transcendent purpose of engineering, to provide authentic situations for students to experience the value of professional skills and have structured opportunities to develop and demonstrate those skills, and to maintain or increase the overall quality of the technical work that students complete on the capstone projects. Based on my 10 years of experience teaching capstone design courses and on relevant engineering education literature, a connected capstone is best when it is an authentic experience that students see as meaningful with real consequences, and where students have a clear sense of the purpose of their project. Reflective learning should by modeled by faculty and practiced by students, especially with respect to the design process, the team process, and the role of professional skills in being a good engineer. Coaching from peers, mentors, and if possible by external professionals should be the primary mode of teaching, and evaluation (grading) should be structured to promote and reward the development and demonstration of professional and interpersonal skills in the context of the design project. The selection and scope of projects, formation and facilitation of teams, and implementation of a grading method that rewards both professional and technical skills are critical to the success of a connected capstone.

The "Designing to Make a Difference" Connected Capstone Experience

The connected capstone described in this chapter is an intensive yearlong experience that has the theme "Designing to Make a Difference" (DMAD), which means that the projects are intended to make a positive impact in the life of a person who does not have the resources to meet his or her need. This mechanical engineering (ME) capstone experience has evolved over a decade from a faculty-controlled, project-based learning experience to a student-oriented, experiential and reflective learning experience, with each evolutionary advancement guided both by the emerging body of educational research and by direct assessment evidence. Student teams of five or six members are formed based on diversity of skills, cognitive styles, and natural team roles. Industrial Advisory Board members, some of whom are graduates who experienced the capstone projects as students, form mentoring relationships with the capstone design teams with a focus on project management and team skills. The design teams complete an entire design process starting from building a relationship with a customer (also referred to as a partner), using interviews and observations to collaboratively define the problem, generating a number of creative alternative solutions, and presenting them to the partner and soliciting feedback that is used in the selection of the final design concept. Each team then designs and builds a working prototype that is delivered to the partner for testing, and the team continues to work with its partner to provide product support so that the partner will have a usable solution before the end of the academic year.

Ohio University's location in the Appalachian region of Ohio lends itself to community-based service learning. The region lacks industry but has a wealth of individuals and groups who have real needs but limited resources to meet them, so local and regional DMAD partnerships are encouraged and are the norm. Individuals with disabilities or physical challenges are common partners. Other projects focus on regionally appropriate technology to promote economic development. Example projects include a pop-nozzle assembly jig to allow more individuals with low physical functioning to participate in an assembly operation at SW Resources (a sheltered workshop employing individuals with disabilities), a community-scale thresher and dehuller to help the Appalachian Staple Food Collaborative in its mission to develop field-to-food systems within communities, and a basketball shooting device to enable a wheelchair-bound high school student with limited arm mobility to participate in wheelchair basketball. Most of the team expenses for the projects are covered through college support and departmental fund-raising, but teams must create a budget and provide justification before funds are allocated. Student reflection reports provide evidence that challenging students to become aware of the needs in the community and giving them freedom to select a community partner and a project helps them make the connection that their engineering skills can be used to serve the needs of society.

The capstone experience is literally connected in several ways. Overlap scheduling is used to allow weekly interactions between the first-quarter freshmen in Introduction to Mechanical Engineering and the seniors in the capstone project so that freshmen see how engineering projects respond to real needs. The seniors also take on a mentoring role and help the freshmen form teams and work on their freshman design projects. Also, a clustered course model links a computer-aided design course and an experimental design lab with the capstone project course so that students learn analytical and experimental skills and immediately apply them.

We assess sudent performance based on a nearly equal weighting of the quality of design work, as demonstrated by the prototype, design reports, and presentations, and the demonstration of professional skills and interpersonal skills using a performance-review model. Project-related evidence of student learning is similar in all capstone courses, but a connected capstone has additional evidence on student behavior in relation to expectations of the profession and student abilities for reflective learning. There are ongoing challenges in developing and maintaining a high-quality assessment process for any course or curriculum, but years of student performance evaluations provide strong evidence that in a connected capstone experience, engineering students demonstrate a strong sense of and capability for developing professional and social skills, even if the majority of their curriculum to date has primarily emphasized the technical skills (Kremer and Burnette, 2008).

Thinking as a Team: Collaboration and Communication

A special focus on collaboration, communication, and other team skills is important in a connected capstone. Interpersonal and social skills are critical, but are often underdeveloped and undervalued by engineering students. Anyone who has been in a group, either in an academic setting or some other setting, has likely experienced how challenging it can be to think and act as a team. Strong-willed individuals (common among engineering students) who are used to controlling their own academic fate often feel uneasy with their success being dependent on others, especially others who are different than them and whom they did not choose to work with as teammates. But engineering and the problem solving and decision making that constitute its practice are highly collaborative activities, so helping students connect their individual engineering skills with team decision making and team productivity is necessary. The ideal, although it is unrealistic, is that the teams in a capstone design experience will not just work together following the guidelines for effective teams (*Harvard Business Review*, 2004), but will also learn together following guidelines for team-based learning (Michaelsen et al., 2002) and collaborative learning (Oakley et al., 2004), and ultimately experience the sense of being a part of a learning organization (Senge, 2000). Even though this ideal is hard to achieve completely, it should remain the goal for the connected capstone. But what is the reality for the students?

A group of engineers that apparently lacks diversity in the standard sense (race, gender, nationality) will nonetheless possess a diversity of skills, intelligence, cognitive styles, learning styles, communication styles, relational styles, decision-making approaches, experiences, perspectives, maturity, biases, opinions, motivations, concern about grades, and so on. Because of the communication challenges presented by diversity, it is essential that an instructor or team facilitator address diversity at the beginning of the team experience in a holistic way and then continue to provide guidance and coaching on effectively using team member diversity throughout the early stages of team development to give student teams at least a chance of reaching their full potential. Experience has shown that it is important to take the time to have team members develop self-awareness of their preferred styles of thinking, learning, communicating, and so on and to develop an appreciation for having different styles represented on a team as essential for it to be more effective than an individual. Based on student feedback, the most effective way to get team members to appreciate the benefit of positive conflict on team performance is to show examples of high-performing diverse teams such as those used in the innovative product development company IDEO.

The example of Team ME1 is instructive on both the benefits and the challenges of diversity. This five-member team had a significant range of cognitive styles, prior experiences, life perspective, and abilities. Several members identified themselves as leaders. During the initial stages of the project where creativity was needed, the team excelled with a wealth of creative concept ideas based on their diverse backgrounds. However, the team got stuck and had difficulty making consensus decisions to move the project forward. Facilitated team meetings and a peer feedback process identified differences in communication styles as the primary cause of their inefficient meetings, inefficient work distribution, and inefficient decision making. This awareness and open discussion of the problem enabled the team to move forward and work together quite successfully. Team members' reflections on the situation showed the power of the learning experience. One member's reflection included the following:

> My team feels that sometimes I don't hear when Joe asks questions. . . . Sometimes I am so focused on writing my design section that everything else is blocked out. . . . It is unintentional and I don't want any of my team members to feel that I am not listening to their ideas. On the other hand, Joe is very creative and has so many ideas.

Some are really crazy and it is hard for me to picture and imagine what he is saying. I struggle with communicating with him in this area simply because I can't picture what he is talking about. Sometimes I am silent for a couple minutes because I am trying to picture his concepts. He has such a creative mind that it is hard to jump in this box sometimes. However, he draws pictures when I tell him I don't understand and this helps open our lines of communication and it is then that I can see what he is saying. . . . I have a lot of plans for growth and development in this area. . . . For starters, I should facilitate more discussion within team meetings. I shouldn't be so focused on what I am working on and open up the lines of communication to all team members. I should take the time to listen when any team member asks a question and really understand what they are saying.

More generally, during the team performance reviews that occur in the first half of the yearlong design process, over 90% of student teams identify communication within the team as the skill that needs the most improvement. Similarly, in the process where team members provide peer coaching to each other, the records of feedback for the past three years show that interpersonal communication is the skill most often mentioned as in need of improvement.

Teachers must also consider the reality of limited time for development of teamwork and collaborative skills. The fact that capstone design teams exist to complete a project provides a constant temptation to emphasize project work over development of skills for team effectiveness. Teams that give in to the temptation become project-obsessed and often abandon planning and scheduling, adopting some variation of a "divide and conquer" approach to tasks and placing short-term progress over long-term effectiveness. These teams eventually experience problems, but this reality cannot be taught; the lesson can only be learned through experience. For many years we emphasized team effectiveness before the teams got fully involved in the project; recent experience, however, has shown that most teams are not ready to take methods for team effectiveness seriously until they have personally experienced the frustration of lack of progress due to ineffective teamwork. Here the example of Team ME2 is instructive. To meet the capstone course requirement, Team ME2 created a set of team operating procedures early in the project, including items like agendas and action items, because they had been presented in class as tools and were included in the example procedures available to the teams. However, in the context of discussing team performance with the instructor after about three months of project work, the team members decided that to improve their performance they needed to get serious about using agendas and action item lists. They had actually forgotten that they had included them in their operating procedures, showing that initially they viewed the tools as course requirements unconnected to their success, but after experiencing team ineffectiveness, they viewed them as connected to how to be an effective engineer.

Using Performance Reviews for Development and Assessment
of Professional Skills

Within an academic setting, grades matter. So even if a capstone project bridges the divide between academia and professional practice, unless the grading system also clearly places value on a diverse range of professional skills the students will primarily pay attention to those things with a significant grade impact. One grading and assessment process that has succeeded in supporting professional skills is an industry-style performance review system that looks at a student's overall performance in professional and technical areas to assign a grade. This is in contrast to a common model where a grade is assigned to a team based largely on team productions like reports, presentations, and prototypes, with peer ratings used to determine team member grades in relation to the team grade. Performance reviews also have the advantage of authenticity, since they are the dominant method for evaluation in the engineering profession. To emphasize this point, members of our Industrial Advisory Board present examples of performance reviews to show students that the methods used to evaluate engineers in industry include a strong emphasis on professional skills. These examples, along with the useful lists of the expected characteristics of engineers (Davis et al., 2005), are a good start in making students aware of the expectations of the profession, but unfortunately these expectations remain largely irrelevant to students until they experience situations where the expected characteristics make a difference—which is exactly what is supposed to happen in the context of a connected capstone experience. However, a project experience by itself is not sufficient. The performance review is a means of focusing student attention on the professional skills through reflective practice. Here the experience of Team ME3 is instructive. This team was working furiously but not making much progress. All team members knew that there was a problem, but most just suffered quietly through the frustration since they did not feel empowered to do anything about it. A team performance review meeting raised the issues and presented the team with several strategies for improving communication, planning, and decision making. After individual reflection and team discussion, team members chose some simple strategies that fit their team dynamic, implemented them, and improved effectiveness. In their end-of-year reflections, they placed higher value on planning and scheduling skills than most other teams, and their comments demonstrated the lessons from reflective learning.

The habits of reflective practice developed in repeated performance evaluations focusing not only on professional skills but also on self-awareness and self-improvement provide a foundation for the type of transformative learning described by Dilworth (2009) that could lead to engineers who truly make a world of difference:

> It takes time and practice to unlock the ability to reflect.... However, once the impasse is breached and reflection starts to occur naturally and routinely, the individual can feel empowered and in control of their own life. That can be a liberating

experience. . . . It can become transformative learning. The individual is elevated to a new plateau of self-awareness. At this point, it becomes what can be called emancipatory learning—throwing off the self-imposed, and frequently externally imposed, chains that have been constraining clear thinking and advance. Reflection in the end is a dialogue with self. (np)

To improve student understanding of professional skills and student acceptance of the performance review model, students should create their own list of essential skills based on their ongoing capstone design project experience, including a textual description of what acceptable performance in those skill areas would look like. For example, in 2006 in the category of "attitude," a description of acceptable performance was "[t]ries to stay positive, looks on the bright side." In repeated surveys nearly 80% of students indicate that their involvement in creating the skills list and describing what the expected behavior would look like is a significant part of the learning experience. In the first few years the performance review model was used, we put significant effort into creating a single performance review document for all teams based on a compilation and synthesis of the individual student input. This effort generated a taxonomy of professional skills for engineering students in a capstone experience, but continued assessments and surveys have shown that having a separate set of expected characteristics for each team provides the same level of motivation and learning with much less effort required for compiling and synthesis (Kremer and Burnette, 2008). Teams may also use an existing list of skills and characteristics, as long as they are required to adapt it to their own situation. For example, in 2009 Team ME4 used a taxonomy of professional skills compiled in 2006 (Kremer and Burnette, 2008) as a starting point and selected seven areas of primary concern in its situation: participation, honesty and integrity, dependability and punctuality, accountability, work ethic, planning, and communication.

A review of the skills and the descriptions provided by students is revealing. Table 4.1 shows common categories of skills included by teams in their performance reviews (Kremer and Burnette, 2008). Nearly every team includes something from each category. The table lists the subcategories in frequency of inclusion by teams.

In addition to their participation in the creation of the required skill lists, students use them to give coaching feedback to their peers, and in a self-evaluation they provide examples or describe experiences that demonstrate their use of several of the professional skills. Finally, all students meet with the professor for a face-to-face performance review to discuss the peer coaching they received, their examples showing their performance in the professional skills identified as important by their team, and a development plan for one professional skill area they need to improve. Students submit a follow-up reflection report after the skills development project is complete, usually near the end of the year. An example statement from a reflection report related to participation and personal development shows the type of reflection that is common:

Table 4.1. Skill Categories Identified as Important by Mechanical Engineering Project Teams

Communication	Teamwork	Effective Meeting Behaviors
a. Interpersonal	a. Dependability/punctuality	a. Staying on task
b. Technical communication	b. Task distribution / does their share of the work	b. Participation
c. Documents work	c. Attitude	c. Facilitation
d. Public speaking	d. Emotional intelligence / interacts well with others	d. Conflict resolution
e. Listens well	e. Supports diversity and specialization	

Character / Personal Foundation	Leadership	Project Management
a. Taking initiative	a. Inspiring and influencing others	a. Organization
b. Work ethic	b. Sharing credit/rewards	b. Planning and scheduling
c. Accountability	c. Coaching, mentoring, empowering others	d. Prioritization
d. Honesty and integrity	d. Leading decision making	e. Business awareness
e. Personal development	e. Vision for project	f. Builds customer loyalty
f. Adaptability / dealing with change	f. Building relationships	

Technical Skills
a. Technical ability
b. Problem solving
c. Creativity

My personal skill development project through the year was to try to do things that maybe I'm not good at already so that I can learn to do these things. I will have to do this once I have a job so avoiding projects that are uncomfortable for me now won't help me NOT avoid them when I'm a part of the work force. I improved this quarter by taking up more of the design report than I did in the past. I wrote a whole section instead of just helping with a few sections. I gained a sense of confidence in my own work. There was a lot riding on the work I was doing because it affects my whole team.

In evaluating the ability of students in a connected capstone to identify professional skills essential to being a good engineer (and specifically things that are important to make their team successful), five years of evidence shows that students are able to develop lists and explanations consistent with, though expressed differently than, those from professional societies. And although on an individual basis few students have a complete picture of the professional skills valued in industry, on a team level and over the course of a yearlong capstone course with multiple peer coaching exchanges and individual performance reviews, all students do demonstrate an awareness of the range of skills and the expectations for professional practice (Kremer and Burnette, 2008).

It is also interesting to look at some of the essential characteristics identified by students. Character and other related characteristics (called a personal foundation) are present in every team's peer feedback and coaching categories when they are asked to identify the characteristics they want to see from all of their team members. Students do not feel comfortable judging others on these foundational characteristics, but they insist on their importance and are comfortable discussing them as part of their personal reflection.

Discussion of Results

The evidence supports the conclusion that the connected capstone is successful in helping students appreciate the purpose of engineering and develop professional skills while doing high-quality technical work. The evidence that students appreciate and connect with the underlying purpose of engineering as service to society includes:

- In student surveys for the graduating class of 2006 nearly 90% of respondents indicated that their understanding of what it means to be a good engineer changed over the course of the project.

- The majority of student feedback at the conclusion of the project and from alumni with between two and five years of work experience lists the capstone project as the most important learning experience in their college career.

- Many students make comments such as the most satisfying experience of my college career was "the look on the face of our customer when we delivered the completed prototype."

There is evidence that students in a connected capstone value and develop a range of skills for professional practice, including the foundational skill of reflective learning for self improvement.

Based on an overall assessment of evidence from written performance review reports that considered the students' ability to describe examples that demonstrate professional skills, to provide an honest rating of themselves and their peers relative to expectations of their profession, to identify a professional skill they need to improve, to plan and conduct a skills development activity, and to reflect on their success in improving that skill, over 95% of the students demonstrate acceptable performance when the process is started early in the year and repeated one or more times, giving students practice with the process and a familiarity with identifying relevant examples of professional-skill-related behavior for themselves and their teammates.

There is evidence that project quality and academic rigor are maintained or improved:

- Based on observations and student feedback, most students in the connected capstone demonstrate high motivation, putting in long hours and showing great perseverance to see projects through to completion.

- The percentage of "successful" projects has increased since moving from competition-style projects to community-centered service learning projects. Importantly, the motivation to put the finishing touches on a project has increased significantly for design teams who establish a personal connection to a customer. It is sometimes said that the last 10% is as hard as the first 90%, so demonstrating this ability to finish a project is an important skill for professional practice.

- Student performance in a connected capstone as judged directly by design report content and presentations is as good as or better than that in a traditional capstone course. There has actually been improved performance with design tools such as Failure Modes and Effects Analysis, likely due to the increased sense of responsibility when delivering a project to a real customer.

- In 2009 one of our capstone design projects, a pop-nozzle assembly jig, was awarded first place (and a $20,000 award) in the National Institute for the Severely Handicapped National Scholar Award for Workplace Innovation and Design. Another project won honorable mention for the team's baked goods storage and display cabinet which the team mounted on a cargo tricycle to enable a local woman with physical challenges to advance her business selling baked goods at a local farmers market.

- In recent program evaluation reports, the Industrial Advisory Board has strongly supported the transition to a connected capstone experience, making special note of its value to students with respect to skills for professional practice.

In order for engineers to make a world of difference, engineering education must lead the way. The principles put forth in *Educating Engineers: Designing for the Future of the Field* provide the vision. The challenge for educators is to work out the best ways to connect concepts and applications, integrate skills and professional practice, and allow students to experience engineering in the world while engaging in reflective practice. With its community-based service learning projects and its use of performance reviews to encourage students to make significant transformations in their own perspectives and skills while working as engineers to make a difference in the life of a customer, the "Designing to Make a Difference" connected capstone experience has proven to be an effective way to implement those educational principles. As an instructor, sharing the experience of students as they engage in reflective practice and make amazing efforts to complete projects that make a real difference has been a transformational experience for me. I echo the words of a former student, who commented just before graduation: "this class has been a journey of personal growth that I have not taken for granted."

References

ABET (Accreditation Board for Engineering and Technology). (2011). *Criteria for Accrediting Engineering Programs: Effective for Reviews during the 2012–2013 Accreditation Cycle.* Baltimore: ABET.

Davis, D.C., Beyerlein, S.W., and Davis, I.T. (2005). Development and Use of an Engineer Profile. *Proceedings of the 2005 American Society for Engineering Education Annual Conference and Exposition.* Portland, OR: ASEE.

Dilworth, R.L. (2009). Creating Opportunities for Reflection in Action Learning: Nine Important Avenues. ITAP International. Available at www.itapintl.com/facultyandresources/articlelibrarymain/creating-opportunities-for-reflection-in-action-learning-nine-important-avenues.html. Accessed Feb. 28, 2009.

Harvard Business Review. (2004). *On Teams That Succeed.* Boston: Harvard Business School Press.

Kremer, G. and Burnette, D. (2008). Using Performance Reviews in Capstone Design Courses for Development and Assessment of Professional Skills. *Proceedings of the 2008 American Society for Engineering Education Annual Conference and Exposition.* Portland, OR: ASEE.

Lord, M. (2010). Not What Students Need. *ASEE PRISM,* Available at http://www.prism-magazine.org/jan10/tt_01.cfm. Accessed Nov. 13, 2012.

Matusovich, H., Streveler, R., and Miller, R. (2009). We Are Teaching Engineering Students What They Need to Know, Aren't We? *Proccedings of the 39th ASEE/IEEE Frontiers in Education Conference.* Piscataway, NJ: IEEE.

Michaelsen, L.K., Knight, A.B., and Fink, L.D. (2002). *Team-Based Learning: A Transformative Use of Small Groups.* Westport, CT: Praeger.

NAE (National Academy of Engineering). (2008). *Changing the Conversation: Messages for Improving Public Understanding of Engineering.* Washington, DC: National Academies Press.

Oakley, B., Felder, R., Brent, R., and Elhajj, I. (2004). Turning Student Groups into Effective Teams. *Journal of Student Centered Learning,* 2 (1): 9–34.

Senge, P. (2000). *Schools That Learn: A Fifth Discipline Fieldbook for Educators, Parents, and Everyone Who Cares about Education.* New York: Doubleday.

Shepherd, S., Macatangay, K., Colby, A., and Sullivan, W. (2009). *Educating Engineers: Designing for the Future of the Field.* San Francisco: Josey-Bass. Downloadable summary available at http://www.carnegiefoundation.org/sites/default/files/publications/elibrary_pdf_769.pdf. Accessed Nov. 13, 2012.

5 Integrative Learning in a Data-Rich Mathematics Classroom

Mike Burke

I wanted to let you know that I was actually moved by your final snapshot. I'm accustomed to thinking critically about language and its sources, but it had never occurred to me that the ability to do mathematical equations might be a feature of (or requirement for!) emancipated thinking. I felt jealous of your students.

FOR THE PAST few years, I have been designing, and assigning, data-based integrative writing assignments in my mathematics classes. Each assignment presents the students with a data set about an important issue. Students are asked to analyze the data mathematically by constructing a mathematical model, and then to use a spreadsheet to implement the model. They are to produce a written paper in which they present their model (with a table and a graph), and then use this work as a basis for any conclusions that they reach. The opening quote is a thoughtful response to a description of this work (in the form of a Keep Toolkit Snapshot) from a bright, well-educated young woman. It is fair to conclude from the quote, I think, that nowhere in her education (an Ivy League education followed by an advanced degree) has this young woman encountered the idea that mathematics has any bearing at all on political, social, or even environmental issues. This quote thus stands as a rather disturbing indictment of us, the higher education mathematics community. What her education lacks (and she is not atypical of our college population) has come to be called quantitative literacy (QL).

There appears to be surprisingly little agreement about what, specifically, constitutes quantitative literacy. In a general discussion of QL, *The Case for Quantitative Literacy*, the Quantitative Literacy Design Team writes that "[q]uantitative literacy is more a habit of mind, an approach to problems that employs and enhances both statistics and mathematics. . . . Unlike mathematics, which is primarily about a Platonic realm of abstract structures, numeracy is often anchored in data derived from and attached to the empirical world" (2001, p. 5). Lynn Steen writes that quantitative literacy is "intertwined with political, scientific, historical or artistic contexts. Here QL adds a crucial dimension of rigor and thoughtfulness to many of the issues commonly addressed in undergraduate education. . . . QL is not a discipline but a literacy, not a set of skills but a habit of mind" (2004, p. 22). Randall Richardson and William McCallum argue that "[q]uantitative literacy cannot be taught by mathematics teachers alone, not because of deficiencies in teaching but because quantitative material must be pervasive in all areas of students' education" (2004, p. 17). For this reason, the writing-across-the-curriculum model seems to offer a promising approach. In this spirit, Steen considers a quantitative literacy course at the college algebra level, perhaps focusing on mathematical modeling rather than on preparation for calculus. Such a course could focus on "data, technology, and quantitative communication. It could serve as the hub of a campus-wide QL program in much the same way as freshman composition anchors writing-across-the-curriculum programs" (2004, p. 39).

I find that through my work I have arrived at much the same conclusions, doubtless by a more circuitous route. My initial motivations were quite modest; I wanted to improve the way I taught functions. About a decade ago, the Harvard Calculus Project popularized the "Rule of Three," the idea that we should routinely examine functions from three different perspectives: numerically (as a table of values), geometrically (as a graph), and analytically (as an equation or formula). As I thought about the "Rule of Three," I realized that a spreadsheet is the natural tool for the study of functions because the user of a spreadsheet uses formulas to build tables, and then the spreadsheet quickly constructs an accurate graph. The beauty of the spreadsheet is that it helps students move from one perspective on a function to another, and thus easily view a particular function from all three. So I began my thinking for this work a number of years ago with this idea: the use of a spreadsheet should help my students come to a deeper understanding of the mathematical concept of a function.

At the same time, I had three other ideas that nicely dovetailed with the use of a spreadsheet. I wanted my students, for motivational purposes, to see some genuine applications of the mathematics they were studying. I wanted to teach through interdisciplinary problems, so that my students would begin to see that knowledge is not constrained by artificial disciplinary boundaries. And I had the conviction that asking my students to write about mathematics would help them clarify their mathematical thoughts. I began designing data-based integrative writing assignments that incorporated all of these ideas. The assignments required use of a spreadsheet to view functions from all three perspectives, the presence of data ensured that the students would see genuine applications, the assignments were interdisciplinary by their very nature, and the final product was a written paper. My goal in all of this was simply to teach a better, more interesting mathematics course.

As I began teaching with these integrative assignments, a funny thing happened. I became very interested in helping my students explore additional, nonmathematical issues. How do we decide what is really true? Can we make these decisions for ourselves, or do we have to rely on experts? What is the role of data and evidence in making decisions? Where do reliable data come from, and how do we treat and interpret the data? What kinds of conclusions can we draw from a given set of data? Given a particular issue, what is the interplay between data, preconceptions, opinion, and belief? What is the proper role of science in making judgments and decisions, and in public policy? These questions are subtle, difficult, and important. And they are rarely addressed in college classrooms, I think; science majors may learn to deal with these issues, but the majority of our students graduate from college without ever giving serious consideration to questions such as these.

Over the course of the past year, I constructed six data-based written exercises for my students in calculus and precalculus courses. The six topics addressed were global warming (based on data about carbon dioxide levels), a historical look at the population of Ireland, radiocarbon dating, global warming again (based on data about the size of the Arctic ice cap), nuclear waste, and world population. My students found these assignments extraordinarily difficult, and I heard, over and over again, comments to the effect that "I have never been asked to do anything like this before." Although I intended these assignments to be interesting supplements to the courses, discussion of the nonmathematical issues described above threatened to hijack my mathematics courses. Students simply found these issues very compelling, and they wanted to think, and talk, about them, virtually every day.

I propose to take a close look at one of the assignments,[1] an assignment on global warming (currently a hot topic—couldn't resist), to provide a more complete picture of the nature of these assignments. The assignment begins with a series of seven quotes; the intention is to simulate the noise that surrounds the issue of global warming in the public arena. A couple of examples:

> The IPCC (Intergovernmental Panel on Climate Change) has provided the world community with first class assessments of the soaring temperatures the world is facing, the devastating impacts of these rises and the ways in which we can try and avoid the worst effects of global warming. We now know climate change is real and the hand of humankind in this warming is becoming clearer and clearer.—Klaus Toepfer, executive director of the United Nations Environment Programme
>
> With all of the hysteria, all of the fear, all of the phony science, could it be that man-made global warming is the greatest hoax ever perpetrated on the American people? It sure sounds like it.—U.S. Sen. James M. Inhofe, Chairman, Senate Committee on Environment and Public Works

The students are then given C. D. Keeling's data on carbon dioxide concentration in the atmosphere, measured on Mauna Loa, at five-year intervals beginning in 1960.[2] I ask students to construct a linear model that fits the data, describing the carbon dioxide level as a function of time. They are to implement the model by using the spreadsheet to construct a table and a graph of the model. Students are then to make a couple of predictions about future levels of carbon dioxide in the atmosphere. When they write their papers, the students are to include a discussion of the derivation of the model (in an appendix to the paper), the table and graph that they constructed, and the predictions that they made. Finally, they are to select a quote from the seven at the start of the assignment and to discuss, using the data and their model as a basis, whether the conclusions they draw from the model are consistent or inconsistent with the selected quote. If they think that the given data and information are insufficient to make such a determination, they are to discuss what additional information they would need to know in order to reach a conclusion. The written assignment emphasizes that they are not to try to prove or to disprove the assertion that global warming is occurring; the information given does not support either conclusion. In particular, students are provided with no information about the link between carbon dioxide levels and global warming, and the students are not to do additional research. The assignment, then, is an exercise that examines the students' ability to draw careful conclusions from the carbon dioxide data and from the additional information given.

Most students do a reasonable job with the mathematical portion of the assignment. They produce a model such as $C(t) = 1.4t + 299$, where t represents the number of years since 1950 and $C(t)$ represents the carbon dioxide concentration, measured in parts per million. They then construct a table and graph: Students are able to embed the discussion of the derivation of the mathematical model, along with the table and graph, into their papers. And they use the model (making use of any of the table, graph, or formula) to make the required predictions.

The remainder of the assignment is more problematic. Many students come to this assignment with very strong opinions about global warming. Some want to write a paper that proves that global warming is not occurring; others want to prove the opposite. They have a difficult time discarding their preconceptions and deciding precisely what conclusions are truly supported by their work with the data. Other students, without strong preconceptions, simply have a very difficult time drawing a careful conclusion consistent with their work with the data; some fail to conclude anything at all in a meaningful way, presenting a couple of contradictory statements and leaving the reader to guess at what they really think. Here are some representative concluding paragraphs. The first two suggest a sense of general confusion; the third offers a more nuanced, thoughtful conclusion:

> Nature has many things in store for us in the short future, but looking back at the past can give us an idea of the general way of things to come. Whoever is right about the global warming debate is irrelevant. Humanity must work to preserve itself. The future has in store great trials for mankind, and maybe we will be better off by trying to preserve our environment, but maybe we will not, nobody really knows. All we can do is look in the past to prepare for the future, the Scripps Institution of Oceanography is taking us the right direction.

> In conclusion I have learned that you cannot predict the future of people's fate as well as the earth's fate. Scientists claim that they found bubbles in the ice that had air in them with low carbon concentration. But the earth has changed and so have

people. People ask "Why study history?" and the answer is and has always been "so that you don't repeat it again." In my view the only thing that will let us know is time, so enjoy everyday of your life. You spend your whole life worrying about the future before you know it you have no time to enjoy life.

The Mauna Loa data shows that the CO_2 concentration over Mauna Loa has been steadily increasing, in a linear fashion, from 1960 to 2000. A model based on this data predicts that carbon dioxide levels may continue to rise, over Mauna Loa, in a similar fashion. What this means is speculative but if we assume that carbon dioxide levels over Mauna Loa [are] representative of global conditions, that carbon dioxide levels [are] responsible for ten percent of the greenhouse effect, and that the greenhouse effect is responsible for global warming, then this could be telling us that global warming may continue to rise, at the same current rate, well into the future. With the proper scientific evidence, or proof, the above statement could become more than just speculative observation; it could be a dire warning.

As I taught the class, I had the general sense that we were making progress in our effort to learn how to work with and interpret data. Class discussion was lively, and students seemed to be beginning to think about the importance of data, and about the idea that their conclusions should follow from their work with the data. And indeed, a careful reading of their papers does show that they are learning how to treat the data mathematically. But as the above paragraphs indicate, interpretation of their work with the data is another matter. They describe their difficulties when I ask them to reflect on the work that they have done. Here are a few of their reflections:

I found it difficult to write about why the statements could not be proved. It's hard to take a passive stance while writing an essay when I have always been taught otherwise.

It was difficult to be as unbiased as possible while writing the paper.

I did not find examining the statements and how they compare to the data particularly challenging. I've found no conclusive evidence (and do not want to jump to conclusions) that man-made CO_2 emissions had any effect on global temperatures. Indeed, the data made no mention of global temperature trends. Even if CO_2 levels and temperatures rose during the same time period, correlation does not imply causation. There is no conclusive evidence that such emissions (which may or may not be man-made) have any significant effect on temperature.

The challenging aspect of examining the data was making sure that I didn't try to prove or disprove anything. I'm used to drawing some sort of definitive conclusion in my essays.

Examining the statements was a new idea for me. I've never done a math paper like that. I found it challenging because you couldn't sound opinionated or seem like you were taking sides.

I tell my students that the central question they are to answer when they are working on their papers is "What do the data tell us?" But their concluding paragraphs and their reflections on their work indicate that many have a difficult time with this. Although my students began to see the value of examining the data, and of letting their work with the data drive their conclusions, many had great difficulty reaching an appropriate conclusion. Some students are very uncomfortable when their data-driven conclusions differ from their preconceived beliefs. Other students do not seem to be able to reach any kind of meaningful conclusion at all. I do think it is our job as teachers to create this kind of discomfort, to push our students to the point where they are seriously examining the data and questioning their beliefs, and to push them to begin to think about how they arrive at their conclusions. It's a struggle for my students.

I began my work with the goal of creating data-based, integrative assignments to supplement the mathematical work the students were doing in my classroom. But the issues raised by these assignments were fascinating, compelling, and complex. I found myself thinking about the role of data in decision making, the care that we must take when we draw conclusions, and the interplay between data, preconceptions, opinion, and belief. I have come to think that the assignments are beginning to encourage in my students a scientific way of thinking, to give my students a new (for them) worldview that incorporates the use of data and evidence and that uses mathematics and careful reasoning to reach conclusions. I think that these lessons about thinking are as important (or perhaps more important) than the mathematics we ordinarily teach.

Where do I go from here? I next want to design, and teach, a mathematics course aimed squarely at quantitative literacy. As Lynn Steen suggests, I can see such a course, taught at the freshman level, as the anchor to a campus-wide QL program. I would make the data-based, integrative assignments the focus of the course, rather than an interesting supplement. I want to have the time, without the pressures of adhering to a set mathematical curriculum, to fully address the issues I encountered last year in response to the integrative assignments, to give those issues the full discussion that they require and deserve.

Notes

1. This assignment was designed in collaboration with my colleague Jean Mach at the College of San Mateo.

2. These data can be found at the website http://cdiac.esd.ornl.gov/trends/co2/sio-mlo.htm .

References

Quantitative Literacy Design Team. (2001). The Case for Quantitative Literacy. In L.A. Steen (ed.), *Mathematics and Democracy: The Case for Quantitative Literacy*, pp 1–22. Washington, DC: Woodrow Wilson National Fellowship Foundation.

Richardson, R.M. and McCallum, W.G. (2004). Embedding QL across the Curriculum. In L.A. Steen (ed.), *Achieving Quantitative Literacy: An Urgent Challenge for Higher Education*, p. 17. Washington, DC: Mathematical Association of America

Steen, L.A. 2004. *Achieving Quantitative Literacy: An Urgent Challenge for Higher Education.* Washington, DC: Mathematical Association of America.

6 Navigating Wormholes

Integrative Learning in a First-Year Field Course

Bettie Higgs

Sᴛᴜᴅᴇɴᴛs ᴏғᴛᴇɴ ᴛᴀʟᴋ of lecturers who are "very good in the field." One day I had the opportunity to assist one such lecturer in leading a group for an afternoon of geological fieldwork. He talked knowledgeably for the duration of the activity, while the students listened. By the end of the day the students had enjoyed the story, questioned nothing, and not recorded anything in notes or sketches. Some time later these students reported that they did not feel confident in the integrative skills associated with scientific fieldwork.

One year later, at the end of a long but rewarding geological field course, I chatted to my first-year students as we set off for home. I was pleased with their comments on how much they had enjoyed the day. I thought I had helped them, with my leading questions, to unfold a clear geological story. So when Mary added, "but I didn't really understand what we were doing," it came as a shock. How many more students "didn't get it"? As it turned out, I had no way of knowing. When I later assessed their field notebooks, they had all written down more or less what I had said. They had sketched what I asked them to sketch. But there was nothing more. Was this my intention? It seems so; the students who had done this well received good marks. But Mary's simple comment has haunted me ever since, and reminds me that understanding is about more than content knowledge, with some associated activity. Both stories suggest that not all activity equates to learning, understanding, and integration. There has to be something more to geoscience field work. What are the secret ingredients and what is worth rewarding?

Lee Shulman's work on signature pedagogies (Shulman, 2005) has helped me to question the purpose of fieldwork as an integrative activity. Are we training students to listen and take good notes, or can we acknowledge what we really value and use this insight to optimize learning during a field course? The argument that has sustained this expensive and time-consuming pedagogy is that the field is where "real" learning occurs; but little evidence of this (except anecdotal) has been forthcoming. When asked to apply their learning, students often cannot demonstrate understanding.

In the United States, Anne Colby et al. (2003) report similar cases where potentially rich experiences have appeared to result in shallow learning. Duncan Hawley (1996), referring to field courses in the United Kingdom, suggests there is a dominance of "the excursion type commonly called the 'Cook's tour' which is characterised by a didactic/instructive teaching approach with passive student interaction" (p. 245). Helen King (1998) worryingly concluded that the majority of institutions fall back on this method as an easy and cheap, though ineffective, option. The situation in Ireland is not dissimilar to that in the United Kingdom. The traditional field trip can sometimes be more akin to a series of lectures in the field, with students writing down whatever the lecturer says, rather than recording their own observations and interpretations. Students who can write quickly and neatly are rewarded when the notebooks are collected for assessment. There may be no opportunity for students on the course to practice being a scientist. These challenges call into question the usefulness of the field experience. Hawley (1998) felt the need to warn us "to be clear about why, how and what styles of field-based teaching and learning are valuable to the learning of geoscience" (p. 10). He asks, "what are the learning processes students go through in field-based learning?" and "do different students learn different things from different types of fieldwork?" (p. 10). He suggests that if we can answer these questions then we can make the most of "being there."

The Broader Context: An Irish Case Study

In University College Cork, Ireland, students apply to enter a broad first-year science program with the intention of following a more specialized program in their second, third, and fourth years. The belief is that this first-year experience will give students a good foundation in all the sciences. The students experience the program as a set of eight discrete science modules in "parallel" disciplines or subdisciplines. There may be little communication between the module coordinators, except to say "students can't seem to transfer their knowledge and skills from one module to another." Indeed the teaching and assessment practices confirm that the courses are self-sufficient and separate. It is up to the students to make the connections between modules if they can, and some do so better than others.

In 2005 a study began to explore the potential of the geoscience "field course" as a way to help first-year students make connections within and between the disciplines they study, *and develop capacities* to deepen, connect, and integrate their learning in a variety of situations. In this "vision of the possible" integrative learners would develop a sense of purpose, ask probing questions to help achieve their learning goals, and fit fragmentary information into a "learning framework." The study deemed fostering these student attributes in first-year science essential.

Design of the Field-Based Course

With these challenges in mind, I, with the help of colleagues, transformed the geoscience field course from a series of lectures-in-the-field into seminars-in-the-field as part of an investigation of how integrative learning could be promoted in first-year science. The design of the course was influenced by the need to:

- decrease didactic teaching, and develop strategies to improve student engagement, student understanding, and student empowerment. This included *developing* key ideas, rather than *presenting* key ideas (Wiggins and McTighe, 1998)
- develop habits of mind, and build capacities, for connection making
- have more frequent ongoing assessments, to see if the students are "getting it"
- bring colleagues along on the journey

The learning outcomes envisioned for the course included explicit reference to integrative learning. For example, students had to be able to "suggest how this work can link to other courses" they had studied. Goals for understanding included that "all things in natural science are connected." To test if the outcomes were achieved, ongoing formative and summative assessments were designed. These assessments generated several types of feedback, including what Dai Hounsell (1987) calls feed-forward. There was no final theory exam.

The course included traditional activities at the heart of all geological field courses, with particular emphasis on students recording their own observations and not those of the leaders. However, the course also created opportunities to connect learning to prior knowledge gained during the year in other science modules, and opportunities to connect science and the community, more intentionally than in the past. These opportunities are our "wormholes." These pathways into parallel universes provide a metaphor for pathways between "parallel" disciplines in the first-year science program. Extending this, a wormhole could equally well connect subsections *within* a science discipline, connect theoretical and field-based science, or connect geoscience and community.

It is interesting to further explore this wormhole metaphor for integrative learning. In 1935 Albert Einstein and Nathan Rosen realized that general relativity allows the existence of "bridges," originally called Einstein-Rosen bridges but later renamed "wormholes" by the physicist John Wheeler. These wormholes act as shortcuts connecting distant regions of space-time. By journeying through a wormhole, you could travel between parallel universes faster than through normal space-time. However, passage is not easy, as wormholes can vary in shape, and have a constriction, a bottleneck or "throat," that necessitates a struggle to get through. Wormholes are unstable places and could collapse and be destroyed when one attempts to enter. Fortunately, calculations show that an advanced civilization might be able to use something physicists call "exotic matter" to prevent the wormhole from closing. In more recent theories, a wormhole can create its own abundant supply of exotic matter. This would allow the wormhole to be big enough and stay open long enough for people to pass through. Once in place, a wormhole is difficult to remove, and if it is navigated successfully, going back is difficult. Further work suggests that a wormhole can have numerous strands, or connecting fibers, and so is more complex than initially thought.

If this is a well-chosen metaphor, it suggests that integrative learning is not necessarily straightforward. It also speaks to the connection making that can take place when intentional opportunities for integrative learning are created, compared to what students may do without these opportunities. The teacher as guide can provide the appropriate exotic matter to aid the students' passage. With time, the enlightened students can themselves enrich the wormhole with exotic matter, allowing it to stay open for the passage of subsequent travelers (peer learning).

Teaching Strategies for Integrative Learning

The teaching strategies employed, to encourage navigation of these wormholes, were influenced by what could be uncovered and discovered by the students themselves. Grant Wiggins and Jay McTighe (1998) warn that "until we feel more comfortable with designing complex learning for uncoverage, and thus [more] familiar with the kinds of instruction needed to develop deeper understandings, our teaching strategies are likely to remain rooted in traditional coverage" (p. 159). That is, the typical assumption is that understanding can be left to chance.

This transformation meant that field course teachers, who traditionally love to "tell the story," had to modify their teaching style, to allow the learner to play a more prominent role. In the words of Larry Malone (2002), the teachers had to "gift the learning to the learner" (p. 3). According to Wiggins and McTighe (1998), "students must come to see that understanding means that they must figure things out, not simply wait for and write down teacher explanations. . . . [T]he learner must make meaning of ideas" (p. 161).

The teaching strategies employed were also influenced by the work of Mary Taylor Huber and Pat Hutchings (2004), who report that first-year seminars, learning communities, problem-based learning, and student self-assessment encourage integrative learning. They describe incorporating learning from a previous course and combining academic and community work, as valid demonstrations of integrative learning. Richard Gale (2004) sums up the ethos of the new fieldwork approach when he describes the seminar as "a pedagogy wherein everyone has a voice and each person's ideas are valued, a venue for exploring varied perspectives, an opportunity to experiment, a way to flesh out skeletal ideas through the challenge of friendly critics" (p. 1). The seminar in the field was thus developed. One benefit of this approach is that it allowed insights into concepts students typically find difficult, and which form blockages to integrative learning. I further develop this theme later.

The field localities traditionally visited did not change, but with integrative learning in mind, I, and colleagues, designed new field-based activities and fine-tuned existing activities. It was possible to integrate many of the pedagogies listed by Huber and Hutchings, and the characteristics advocated by Gale, in this new module. Careful development of a field workbook for students assisted the integration; the workbook acted as a guide, a resource, and a repository. The science content and concepts became more challenging as the course proceeded, and the purpose of learning moved toward deeper understanding, less dependence on the teachers, and more student autonomy. Active learning communities evolved naturally. The teachers facilitated interactions and dialogue, but the students had to engage in order to learn. The nature of the three-day residential field course, building on the campus preparatory work, allowed students to develop some of the narrative and build their own meaning, guided by the experts.

As part of the assessment, the course designers attempted to reward attributes that indicated "good attitude to learning." Field-course leaders have long recognized that students whose participation is particularly beneficial to the group of learners don't always get the best marks, and in extreme cases can even fail because a notebook is incomplete. A rubric related to Benjamin Bloom's affective domain was modified, to provide criteria such as "attentive; asks for clarification; volunteers; demonstrates commitment to improving," through several categories up to "concerned with bringing the different pieces of learning together; resolving conflicts in knowledge; sees the need for planning; proposes; revises; solves, internalises," and so on. The mark awarded to each student was only 10% of the module. Though a blunt instrument for a complex quality, it began to redress the emphasis in student assessment, and encourage students to enter the wormholes.

Evidence of Students' Integrative Learning

Based on observation and discussion with colleagues, it was clear that great learning went on in this field course. The engagement of students was far greater than usual. The voices of students were heard more often. The faces of the students were brighter. Students' work was of a higher quality. For most teachers in most courses this informal assessment, together with a standard "how did I do?" questionnaire, is as far as course evaluation goes. However, a scholarship of teaching and learning approach requires the intentional collection and analysis of evidence on how well, or otherwise, the teaching strategies worked, and how they affected student learning.

In this study, we divided the 80 student collaborators into two cohorts of 40, and taught and studied the field course twice. Each group of 40 was subdivided into five small working groups. We used a wide variety of data-collection methods to capture the complexity of student learning in the field setting, and quantitative and qualitative methods where appropriate to analyze leader observations, talk between students (peer learning), talk between students and leaders, written work, field sketches, and video and sound recordings. The nature of the activities, and the presence of several leaders, allowed me to react to "spontaneous integrative moments" and occasionally step back and "observe the action," though these observations could never be as completely objective as those of a real outsider. The results of this study have been documented elsewhere (Higgs, 2007, 2008) but I will make a few points here.

Campus-based preparation: Prior to the field-based experience, each small group carried out a preparatory research project on a topic relevant to the field area. They chose the topics from a list especially designed to bring in, or integrate, another "angle" or area of science not traditionally included in the course. This project encouraged students to get to know each other before working together in the field. The assessment criteria indicated that the final written report should show integration of preparatory research (campus-based) with the new insights gained during the subsequent experiences in the field. A 90-minute interactive activity, run by the Library Information Literacy group, honed research skills, ensuring cocurricular activities connected in an authentic way.

We asked each group to present two brief progress reports to the whole class, one before going into the field and one during the residential field component. This process opened up dialogue between the student groups, and between students and leaders, and allowed students to assess their own performance and move learning forward. In this way all students received formative feedback, without 80 pieces of work having to be read and commented on. Students had the opportunity to hear each other's views, and gain a shared understanding of what was required. For example, one group had communicated with the geological curator at the Ulster Museum, thus connecting with the "real world." This caused other groups to ask themselves "who should we contact?" This feedback was timely, and allowed students to act on it and improve work as it developed during the course. It gave students a sense of autonomy, and addressed the student concern that feedback comes too late when given out at the end of a module (Hounsell, 1987; Samball et al., 2007). Each group was asked to record the *process* of carrying out the project. In the final report the descriptions of process showed a high level of engagement. Having to write these descriptions encouraged students to show insight into their own learning and acted as a catalyst for metacognition.

How did this group research project contribute to integrative learning? The student engagement in this part of the course exceeded my expectations. All the group projects contained evidence of connection making between campus- and field-based study, with others in their group, with other disciplines and with prior knowledge. For example, a project on landslides moved from the geological considerations to the role of vegetation in controlling landslides, and from the affect on local communities to the engineering solutions adopted on the north coast of Ireland. In a separate project, students researching groundwater considered the potential hazard of pollution from local agriculture and industry in Northern Ireland. The work clearly showed that first-year students are well able to carry out collaborative research projects, and indeed that having a purpose for the work (visiting the area of study and reporting the information to the rest of the large group) was a motivator to get them to engage. Students' final reflections included: "In the end, it was an all round belief that having a particular angle to focus on in relation to County Antrim was both important and beneficial. Our field studies were helped immensely from this project and the acquired skills will aid us in the future. A lot was put into this project and we all got a lot out of it." Describing process, and reflecting, was making students more aware of their learning, and building capacity to integrate learning. We should continue to build on the complex step of integrating individuals' work within the group, so that the whole is greater than the sum of the parts, in future years.

This group collaborative work began to build learning communities that made the residential field course, as a whole, more effective. Reflections indicated that students were beginning to understand how they learn in collaboration. By the end of the course students began to realize that the teacher is not the sole supplier of feedback and that they received valuable feedback from their peers. Recognizing and using valid feedback, such as peer discussion, is an important characteristic of the integrative learner, and counters the danger of keeping good learning a secret.

Field-based observations: A field workbook was designed to guide students in the recording of their own observations in the field. Field sketches were encouraged as a means of integrating potentially significant information for subsequent interpretation. Pulling these strands together to aid understanding is integrative. Certain questions within the workbook acted as an introduction to reflective journaling, and a safe way to begin to build capacity to integrate and reflect on ways of learning.

An example of a wormhole activity was measuring the pH of the soil overlying the bedrock type at several sites. Students later compared the values obtained, and observed that they showed marked differences from one rock type to another. This activity had not been incorporated into the course before, even though each measurement takes only a minute or two. It provided rich potential connections with other subjects of scientific debate. For example, some students discussed the changes in vegetation, and others discussed the chemistry of soil pH, which led to a discussion of bedrock weathering and carbon dioxide levels in the atmosphere. These topics have relevance to subjects of interest to society such as climate change, thus providing potential links between ancient rocks and topical issues of the day.

At first there was some resistance to this idea of thinking outside the discipline. Only a small number of students found this a safe place to go. Subsequent examples of wormhole activities provided more discussion and connection making. One example involved a shallow geophysical survey related to a land-development proposal, where students had to connect geology and geomorphology to concepts they had encountered in physics. This led to discussion of applied geology, the needs of the local community, and potential employment opportunities—always of interest to students. The discussion prompted postgraduate leaders to link examples from their own research. Here the clear *purpose* of the activity was a key factor to student engagement.

As the field course progressed, students (and leaders) became more comfortable with the idea that all things are connected. Students realized that they did have prior knowledge that could be brought to bear in discussions. Other wormhole activities linked prior knowledge to fieldwork in a revision exercise at Cushendun; the history of science with present-day understandings of field evidence at Portrush (a key site for the "origin of igneous rock" controversy in the 18th century); rock properties, landslides, and local community at Garron Point; geological time and present-day juxtaposition of strata at Murlough Bay.

The potential of these wormhole activities lies in the fact that *everything* is connected by some pathway, making student learning unpredictable. We can liken pathways to the neuronal networks and connections of the brain, giving each of us our own mind (Greenfield, 2004). It suggests that *all* robust learning is integrative, and relies heavily on previous experiences. So what were our wormhole activities representing? In this study, the wormholes encouraged intentional integrative learning and connection making between "parallel" science disciplines and subdisciplines, and between campus-based science and the "real" world.

The evening synthesis and discussion sessions: We introduced evening seminars which were designed to play a major role in helping students integrate pieces of learning and build understanding. Synthesis activities mirrored the work of real field geologists, where all information is recorded in a geological column or a geological cross-section. These activities help geologists connect the various pieces of information gained through field study. They are thus authentic integrative exercises with a clear purpose. Practicing "geologist as detective," the students pieced together the clues they found during the day. Video evidence collected during the day added significantly to the student and leader engagement in the evening sessions. They brought context back to the whole group, allowing the integration of small group and whole group discussion (Higgs, 2003). It was clear that the course was less about lecturer performance and more about student learning.

In the subsequent focus group all students agreed that "we needed these evening sessions to pull things together" and one student reported, "At White Rocks Bay, when discussing the different methods that possibly formed it—the discussion in the evening with everyone's view—that was my 'ah-hah' moment."

Questions for students about their own learning were woven into the disciplinary program to help build students' capacity to be integrative learners. For example, during the third evening session, after a challenging daytime activity, student groups were asked to defend their interpretation of the evidence they collected. Groups had usually come to a consensus, but listened to the evidence and arguments of other groups, and became persuaded to a greater or lesser degree by them. Questions in the workbook asked students "What did you learn from others in your group?" and "What did you learn from other groups?" When students were asked to explain the arguments of a peer, they demonstrated their own ability to explain scientific concepts and to integrate them with their own views. When students see their peers viewing the world around them, and becoming engaged and vocal, they tend to want to join in. Students also learned that in science there is not always an easily identifiable right answer.

How well were students integrating their learning? We asked students simply to consider "what questions remain?" Six levels of questioning resulted, which revealed much about each student's attitude toward learning. At the simplest level students asked, "which is the right hypothesis?" At the more complex levels, students questioned potential survey methods that could be used to discriminate between hypotheses. An interesting point is that all the students could have formulated the more complex questions. They all had the prior knowledge. It requires an attitude—a frame of mind—to push that little bit further, to navigate the wormhole, and to try to resolve the conflicts in knowledge.

Levels of Integrative Learning: Implications

The course at the center of this study was designed, intentionally, to maximize the opportunities for integrative learning, and to help students develop their capacities to become integrative thinkers and learners. The students had the key role in the learning that took place. As they engaged with and navigated the wormholes, and undertook the formative and summative assignments, we as teachers could recognize "performances of integration" and divide them into three broad categories:

> students can make the connections themselves,
> leaders or peers "provoke" connection making,
> leaders or peers must explicitly point out the connections.

The evidence shows that a student does not belong to one category all the time, but depending on the complexity of the connection, and the student's attitude, motivation, or inclination at any one time, the student moves between categories. However, some students will be in the first category more often than others. We are all born with high learning potential, and yet the evidence shows that not *all* students will go all the way to make meaningful connections, even when multiple opportunities are provided. The concern should then be about the level of integration that takes place. Can we encourage students to operate more often at the higher levels?

These observations map very well onto Ron Ritchhart's internal-external model of dispositions (Ritchhart, 2002). He describes dispositional action, where patterns of behavior are self-initiated, intentional, and consciously controlled; assisted action where patterns of behavior rely on a combination of internal and external triggers; and coerced action, where patterns of behavior occur only in the presence of external support and motivation. His model allows movement back and forth between these three dispositions. Ritchhart's findings concur with those described in this study in that it is not always student ability that is lacking, but rather the inclination to perform. He calls this the ability-action gap. As Ritchhart says, just having the ability is not enough. The student must use that ability to demonstrate integrative learning.

The field course provides geoscience students with authentic opportunities to practice and reinforce important concepts and connections. Why then do students not always make the effort to succeed? Revisiting our metaphor, and focusing on the third category of performances of integration, how can we encourage students to enter the wormhole and attempt to navigate through it? Perhaps students hesitate because the wormhole is an alien and uncomfortable place to be, and requires a struggle to pass through. Yet navigating a wormhole successfully can be transformative, create new knowledge and understanding, and allow further learning to proceed. The question is, what will tempt students to engage, and what kind of exotic matter will students require in order to succeed? A better understanding of this issue can help us craft the opportunities to connect, and encourage students to close that ability-action gap.

In an attempt to further understand the integrative learning process, which creates new knowledge that is more than the sum of the parts, it seems reasonable to view the process from multiple perspectives. How can we encourage a student to enter into the process? What are the barriers that may deter that student from entering the wormhole? Once inside, what is the nature of the struggle?

Border Crossings and Integrative Learning

Olugbemiro Jegede and Glen Aikenhead (2004) offer some insight into why a small number of students don't "enter the wormhole." They revisit and develop the concept of "border crossings" to give us insight into the reality that students live and work in more than one domain or culture. They found that "for many learners conventional science seems disconnected from practical ends" and not every learner has the capacity to resolve conflicts between the subcultures of families, peer groups, the broader community, and university science. Their work deepens the language of integrative learning, and is central to an understanding of the process. It articulates the potential for smooth, manageable, or rough border crossings between the domains that affect student learning. At one extreme, conflicting views (e.g., campus science versus real world) are held separately and do not interact. They are compartmentalized. William Cobern (1996) referred to this as cognitive apartheid. At this extreme, integrative learning is not taking place. At the other extreme, conflicts interact and are resolved in some manner, and students make meaningful connections between university science and their broader community. This outcome indicates that integrative learning has taken place. Between the two extremes are varying degrees of interaction. This aligns to some extent with the findings reported above that there are levels of integrative learning, with students moving back and forth between the levels.

Students may not turn ability into engagement and action, when the culture of science is at odds with a learner's lifeworld. They may feel alienated and discouraged as learners. Jegede and Aikenhead (2004) say learners can develop "clever ways to pass their science courses without learning the content in a meaningful way" (p. 156). J. O.

Larson (1995) describes Fatima's Rules, where students develop coping mechanisms such as silence and evasiveness. The result is not meaningful learning, but communicative competence (Jegede and Aikenhead, 2004). The concern is that these students may not be poor learners, but they learn how to learn sufficiently well to succeed. If there is no reward for navigating a wormhole, then students may not expend the required time and effort to engage with the opportunity.

What can teachers do to help? Huber and Hutchings (2004) advocate intentional teaching to assist students to develop capacities to integrate their learning between academic work and the various aspects of the real world, including community. Understanding the nature of border crossings can inform the design of activities to promote integrative learning. Intentional teaching can build bridges between the worldview of science and the worldview of learners, and smooth the border crossings. First-year science could be taught from the angle of community concerns, such as geohazards, resource exploitation, groundwater pollution, coastal erosion, environmental management, ethics, and citizen science. Teachers can choose teaching strategies to help students identify conflicts, explore them, and feel secure with them. What type of exotic matter do learners need to negotiate conflicts and become integrative learners? Some require a high degree of guidance, incorporating a range of teaching strategies. Others may only need incentives and ideas that create the need to know science. During a residential field course, there is time and space to adapt the degree of guidance and provide the appropriate wormholes between the domains.

Providing Exotic Matter: Insights from "Troublesome Knowledge" and "Threshold Concepts"

All things are connected in some way, and an infinite number of connections could be encouraged. When a student is willing to engage and enter the wormhole, how should we focus our efforts? A number of authors can offer us insights. David Perkins (1999) suggests that there are some concepts that are difficult for students to grasp owing to their counterintuitive, seemingly alien or complex nature. He refers to these concepts as troublesome knowledge. He also introduces the concept of breakthrough thinking (Perkins, 2000), where the learner struggles to make sense of messy data before a "light goes on." Building on this work, and using examples mainly from economics, mathematics, and science, Erik Meyer and Ray Land (2003, 2005, 2006a, 2006b) found that certain concepts were held to be central to the mastery of a subject. They called them threshold concepts. The idea that there are threshold concepts in each discipline came out of a national research project in the United Kingdom (Enhancing Teaching-Learning Environments in Undergraduate Courses, http://www.tlrp.org). Their work aligns well with this study in that grasping a threshold concept is integrative—it can transform learning by exposing the previously hidden interrelatedness of phenomena. Grasping a threshold concept is probably irreversible; if it is robustly understood, the learner is unlikely to forget it.

Meyer and Land (2005, 2006a) refer to threshold concepts as "conceptual gateways" or "portals" leading to a previously inaccessible way of thinking about something. In this visualization the threshold concept can be likened to a wormhole the learner must pass through to experience the connected science. Threshold concepts are likely to involve forms of troublesome knowledge (Perkins, 2006), just as wormholes require the learner to struggle to get through the bottleneck.

This work suggests that if wormhole activities are designed to focus on a threshold concept, then successful navigation represents significant integrative learning. In grasping a threshold concept, the student can go from naïve to deeper understanding within the discipline. These ideas have implications for curriculum design and for pedagogy. Glynis Cousins (2006) believes "in contrast to transmitting vast amounts of knowledge which students must absorb and reproduce, a focus on threshold concepts enables teachers to make refined decisions about what is fundamental to a grasp of the subject they are teaching. It is a 'less is more' approach to curriculum design" (p. 5).

In this study, we asked experienced teachers and students, "what do students find difficult in geoscience?" The consensus was that an appreciation of the enormous span of geological time is crucial but can be troublesome, particularly in imagining the scene "at the time," when surroundings were different, and processes might have been occurring at a different rate to those in the present day. In addition, visualizing in three dimensions is vital but proves troublesome. Geologists often work with two-dimensional information, but must interpret the "patterns" to construct the three-dimensional reality. Many students find this difficult, and assume this is their own innate deficiency that they can do nothing about. Both appreciation of geological time and working in three dimensions successfully require threshold concepts to be grasped. They give an indication of where wormhole opportunities should be concentrated in the future.

Interestingly, the work of Meyer and Land (2006b) reinforces the wormhole metaphor. Their work indicates that the process of trying to pass through a wormhole may not be easy. They suggest learning may involve spending extended time in liminal space. This is an unstable space—oscillating between old and emergent understandings—and may be associated with feelings of confusion, anxiety, and discomfort. This conception aligns with the constriction or throat as students try to pass through the wormhole. It is here that students may need guidance or exotic matter as they become engaged in the process of integrative learning and mastery.

The residential nature and structure of the field course allows a recursiveness, so that students can begin to see patterns and themes, and relate these to ideas, as advocated by Susan Greenfield (2004). This is how meaning is built. Cousins (2006) suggests that grasping a threshold (key integrative) concept may involve leaving an old belief system behind, and can require a difficult repositioning for students. The findings of this study agree with Cousins that students must take risks to make troublesome connections. As teachers we must encourage this risk taking, and explicitly reward it. This incentive will encourage students to navigate the wormholes.

Building on this idea, there may be threshold concepts, the mastery of which is required to make significant connections between disciplines, between campus and the world outside, and between science and the community. That is, interdisciplinary threshold concepts must also exist. Once identified and negotiated, they could be integrative, transformative, and irreversible, and could enable decisions on what is fundamental to a grasp of the interconnectedness of disciplines and domains. If we align this approach with a learning-outcomes approach, each learning outcome would be associated with a threshold concept—what the student will master as a result of taking the course. These insights could be used to design significant wormhole activities, and leave less troublesome connections to be addressed by students' self-directed study.

How can we design curricula to encourage students to enter a wormhole, a liminal space, or a zone of proximal development (the latter described by Lev Vygotsky (1978), where the student can be effectively aided by the teacher)—and turn ability into action? We must alert ourselves to the transformative points where intentional teaching is required. Then we must allow students to spend some time in the liminal space, or "hang out" partway through the wormhole, while questions are answered. As Cousins (2006) says, this space should allow recursiveness and excursiveness, that is, looping back on the conceptual material as opposed to a simplistic linear approach. This assertion would agree with the findings of Greenfield (2004) that multiple connections build robust learning. There is no easy navigation through the wormhole. Significant learning takes place when students struggle with "messy data" to develop robust understanding. Perkins (2000) highlighted this "messy" stage as a precursor to breakthrough thinking.

Feedback as Exotic Matter

To succeed, students must be aware of the opportunities that exist, and must know what action is appropriate. Feedback is crucial and may in the end be the key exotic matter students will require to successfully navigate the wormholes. Therefore it must be focused, appropriate, and timely.

Kay Samball et al. (2007) were concerned that students and teachers did not make use of the full range of feedback that was available to them, and so they have introduced ways to encourage students to develop an understanding of the range and types of feedback that can help them in their learning, including teacher-to-student and peer-to-peer feedback, assessment criteria, and rubrics (see FAST Formative Assessment in Science Teaching, http://www.open.ac.uk/science/fdrl/).

They see feedback as potentially integrative. In an attempt to visualize their proposed model of student learning, they use the metaphor of an electrical circuit board, with switches (gaps) that must be closed for the current to flow, and connections to be made. The catalyst that closes the switches is feedback. New ways of seeing material are shown as light bulbs which come on when new connections are made. Student learning, and the ongoing complex development of understanding, can take a variety of possible paths on the circuit board (compare Greenfield's 2004 neuronal connections, with unique pathways assuring that learning is different for each individual). At the simplest level, a switch may be closed by a student simply getting the right piece of information as feedback. This work suggests that a greater appreciation, by both students and leaders, of the broad range of feedback opportunities could aid integrative learning in the future. The authors insist that it is the student as "self-assessor" who must do the work, recognizing and taking ownership of the feedback, to close the switches effectively, and allow the learning journey to progress. The authors see feedback (exotic matter) as the specific module, with its assignments, activities, protocols, briefings, teaching strategies, leaders, and peers all contributing to connection making.

Paul Black and Dylan William (1998) argue that effective feedback has three elements: recognition of the desired goal (or purpose), evidence about the present position, and some understanding of a way to close the gap between the two. This gives us another insight into the student ability-action gap. Perhaps some students in the field course are not seeing the goal, or are not seeing where they are in relation to the goal, and so they do not attempt to close the gap. The switch stays open. Black and William maintain that all three elements of feedback must be understood to some degree before learners take action to improve learning. Samball et al. (2007) conclude, "in our view the potential of informal feedback, especially the conversations that go on, often unseen, amongst students and tutors, are probably amongst the most influential ways of helping students, make connections for themselves" (p. 12). This informal feedback is one of the key strengths of residential field-based learning.

What Can Teachers Do?

As teachers we are students of design and strategy for better learning. We need to understand how students learn and how our actions enable or inhibit learning. This understanding can help us facilitate closure of the student ability-action gap. In addition, leaders who try to promote a particular disposition may not succeed if they do not themselves have that disposition. For students to engage in integrative learning, their teachers must model integrative learning. This is not something that comes naturally to all, and it benefits from continual development and sharing of ideas.

In this study, we chose initially the more obvious connections to design wormhole activities. As the study progressed, evidence from students, and from experienced field leaders, helped to define the threshold concepts required for mastery of field-based learning in geology. In the future, program and course learning outcomes could be aligned with threshold concepts, to encourage students to engage with them. Wormholes can be designed and put in place to assist in the mastery of these concepts. The students can be encouraged to discover less troublesome concepts by self-directed study, integrating them into a robust framework of understanding.

We took a close look at some exotic matter and found that teachers themselves may contribute to the difficulty of navigation. When experienced field leaders were asked, "What causes difficulty for students' learning?," they admitted that as leaders on field trips we commonly point things out that cannot be seen with the naked eye. We are using information that is common knowledge to the experienced geologist, but is still at the level of theory to the student. This practice can lead students to become dependent on the teacher, to lack confidence, and to consider themselves "not very good in the field." So teachers can create troublesome knowledge. Teachers must build student confidence by encouraging a culture of questioning. Students need to know that questioning is key to their learning. As in Vygotsky's zone of proximal development, what the teacher does to assist the learner navigate the wormhole is crucial.

In some cases leaders may give a lecture in the field, and may present interpretations as facts. The student writes down whatever the leader says. The leader is the only authority figure, and the student does not build up confidence in "doing fieldwork." The student believes there should be a right answer and does not grasp the concept of provisional interpretation and uncertainty. If the teacher does not articulate the concept, then the student may find knowledge troublesome and become despondent. In this scenario students are unlikely to build the capacity to be integrative thinkers and learners.

Returning to curriculum design, and to further analyze what worked, and what can be improved, the key findings of the Integrative Learning Project carried out by the Carnegie Foundation for the Advancement of Teaching and the AAC&U (Huber, 2006; Hutchings, 2006; Gale, 2006; Miller, 2006) were mapped against elements of this field-based course. This process documented the integrative opportunities, which can often seem nebulous, and highlighted some gaps in the geoscience program that may have hindered integrative learning in the past. The results suggested that we can improve integrative learning by increasing the links and opportunities to connect over the entire four-year undergraduate program, writing specific program learning outcomes for integrative learning, and assisting staff in curriculum design, assessment, and pedagogies that have proved to promote integrative learning.

The concept of integrative learning and the wormhole metaphor have helped turn a lens on the field course as a signature pedagogy in the geosciences. Providing students with opportunities to make connections between separate pieces of learning has helped to build their capacities to be integrative learners. Consideration of the importance of feedback, and the idea that there are threshold concepts to be grasped, has helped to give insights into the blockages that may prevent some students from linking theory to practice, linking laboratory work to the field experience, linking academic and real-world learning, and linking naïve to deeper understandings. This idea brings richness and clarity to the understanding of integrative learning itself, and helps us to answer the often asked questions "what is integrative learning, what does it look like and how can it be assessed?"

This exploration has highlighted another characteristic of integrative learning. Integrative learning can vary in the "type" or nature of connection and the time required to make the connection. The learner may not attempt the connection, may try and not succeed, or may struggle long enough to navigate successfully.

Spending time in a wormhole, which is by definition an unstable space, can be a dangerous occupation for students. They will need their own energy to enter the wormhole and begin the struggle, and may need exotic matter (assistance from a peer or guidance from a teacher) to keep the wormhole open and allow safe passage. A complication arises here, which is that common knowledge is different for each student, and may be much different from common knowledge of the teacher. The threshold between common knowledge and theory moves as experience is gained. So theoretically, thresholds are in different places for each student. Each mind is unique, and what is an "ah-hah" moment to one may be commonplace or a linear progression to another. Each student may need different types and amounts of exotic matter.

In the end, if a connection is important enough, the learner must be allowed time to visit and revisit it in multiple ways, and feel safe that mistakes and confusion are tolerated, on route to understanding. An awareness of this tolerance, and even expectation, may encourage more students to take the plunge and turn their ability into action.

References

Black, P. and William, D. (1998). Inside the Black Box: Raising Standards through Classroom Assessment. *Phi Delta Kappan*, 80: 139–148.

Cobern, W.W . (1996). Worldview Theory and Conceptual Change in Science Education. *Science Education* 80 (5): 579–610.

Colby, A., Ehrlich, T., Beaumont, E., and Stephens, J. (2003). *Educating Citizens: Preparing America's Undergraduates for Lives of Moral and Civic Responsibility.* Carnegie Foundation for the Advancement of Teaching. San Francisco: Jossey-Bass.

Cousin, G. (2006). An Introduction to Threshold Concepts. *Planet*, 17: 4–5. Available at http://neillthew.typepad.com/files/threshold-concepts-1.pdf. Accessed Nov. 21, 2012.

Gale, R. (2004). The Magic of Learning from Each Other. Carnegie Foundation for the Advancement of Teaching Carnegie Perspectives, pp. 1–2. Available at http://www.carnegiefoundation.org/perspectives/magic-learning-each-other. Accessed Nov. 18, 2012.

Gale, R. (2006). *Fostering Integrative Learning through Pedagogy.* Stanford, CA: Carnegie Foundation for the Advancement of Teaching. Available at http://www.carnegiefoundation.org/files/elibrary/integrativelearning/uploads/pedagogy copy.pdf. Accessed June 15, 2009.

Greenfield, S. (2004). The Child. In E. Scanlon, P. Murphy, J. Thomas, and E. Whitelegg (eds.), *Reconsidering Science Learning*, pp. 41–57. London: Routledge Falmer.

Hawley, D. (1996). Changing Approaches to Teaching Earth-Science Fieldwork. In D.A.V. Stow and G.H.J. McCall (eds.), *Geoscience Education and Training: In Schools, Universities, for Industry and Public Awareness*, pp. 243–253. Rotterdam: A.A. Balkema.

Hawley, D. (1998). Key Skills and Geosciences Fieldwork: Inseparable Partners in a Total Learning Environment. In H. King (ed.), *Examples of Good Practice in Earth Science Learning and Teaching: Fieldwork*, pp. 9–16. UK Geosciences Fieldwork Symposium: Proceedings. Plymouth, UK: Higher Education Academy, GEES Subject Centre.

Higgs, B. (2003). Virtual Fieldtrips: Bringing Context into the Classroom. Report of Innovative Teaching and Learning Project, University College Cork.

Higgs, B. (2007). Promoting Integrative Learning in a First Year Earth Science Field Course. MA thesis, Department of Geology, University College Cork.

Higgs, B. (2008). Promoting Integrative Learning in First-Year Science. In B. Higgs and M. McCarthy (eds.), *Emerging Issues II: The Changing Roles and Identities of Teachers and Learners in Higher Education*, pp. 37–50. Cork, Ireland: National Academy for the Intergration of Research, Teaching and Learning (NAIRTL).

Hounsell, D. (1987). Essay Writing and the Quality of Feedback. In J.T.E. Richardson, M.W. Eysenck, and D. Warren-Piper (eds.), *Student Learning: Research in Education and Cognitive Psychology*, Milton Keynes, UK: SRHE/Open University.

Huber, M.T. (2006). Fostering Integrative Learning through the Curriculum. In *Report of the Integrative Learning Project*, pp. 1–12. Carnegie Foundation, Stanford, CA. Available at http://www.carnegie.org/elibrary/integrativelearning. Accessed Nov. 18, 2012.

Huber, M.T. and Hutchings, P. (2004). *Integrative Learning Mapping the Terrain*. Washington, DC: AAC&U; Stanford, CA: Carnegie Foundation for the Advancement of Teaching.

Hutchings, P. (2006). *Fostering Integrative Learning through Faculty Development*. Stanford, CA: Carnegie Foundation for the Advancement of Teaching. Available at http://gallery. carnegiefoundation.org/ilp/uploads/facultydevelopment_copy.pdf. Accessed Nov. 21, 2012.

Jegede, O.J. and Aikenhead G.S. (2004). Transcending Cultural Borders: Implications for Science Teaching. In E. Scanlon, P. Murphy, J. Thomas, and E. Whitelegg (eds.), *Reconsidering Science Learning*, pp. 153–175. London: Routledge Falmer.

King, H. (1998). The UK Geosciences Fieldwork Symposium. In H. King (ed.), *Examples of Good Practice in Earth Science Learning and Teaching: Fieldwork*, pp. 2–8. UK Geosciences Fieldwork Symposium: Proceedings. Plymouth, UK: Higher Education Academy, GEES Subject Centre.

Larson, J.O. (1995). Fatima's Rules and Other Elements of an Unintended Chemistry Curriculum. Paper presented at the American Educational Research Association Annual Meeting, San Francisco, April 1995.

Malone, L. (2002). Peer Critical Learning. In *Final Report*, pp. 1–3. CASTL. Carnegie Foundation for the Advancement of Teaching. Available at http://www.carnegiefoundation.org/. Accessed July 9, 2009.

Meyer, J.H.F. and Land, R. (2003). Threshold Concepts and Troublesome Knowledge: Linkages to Ways of Thinking and Practicing within the Disciplines. In C. Rust (ed), *Improving Student Learning Theory and Practice—10 Years On*, pp. 412–424. Oxford: Oxford Cemtre for Staff and Learning Development, Oxford Brookes University.

Meyer, J.H.F. and Land, R. (2005). Threshold Concepts and Troublesome Knowledge (2): Epistemological Considerations and a Conceptual Framework for Teaching and Learning. *Higher Education*, 49 (3): 373–388.

Meyer, J.H.F., and Land, R. (2006a). Threshold Concepts and Troublesome Knowledge: An Introduction. In J.H.F. Meyer and R. Land (eds.), *Overcoming Barriers to Student Learning: Threshold Concepts and Troublesome Knowledge*, pp. 3–18. London: Routledge.

Meyer, J.H.F. and Land, R. (2006b). Threshold Concepts and Troublesome Knowledge: Issues of liminality. In J.H.F. Meyer and R. Land (eds.), *Overcoming Barriers to Student Learning: Threshold Concepts and Troublesome Knowledge*, pp. 19–32. London: Routledge.

Miller, R. (2006). *Fostering Integrative Learning through Assessment.* Stanford CA: Carnegie Foundation for the Advancement of Teaching. Available at http://gallery.carnegiefoundation.org/ilp/uploads/assessment_copy1.pdf. Accessed Nov. 21, 2012.

Perkins, D. (1999). The Many Faces of Constructivism. *Educational Leadership*, 57 (3): 6–11.

Perkins, D. (2000). *The Eureka Effect: The Art and Logic of Breakthrough Thinking.* New York: Norton.

Perkins, D. (2006) Constructivism and Troublesome Knowledge. In J.H.F. Meyer and R. Land (eds.), *Overcoming Barriers to Student Learning: Threshold Concepts and Troublesome Knowledge*, pp. 33–47. London: Routledge.

Ritchhart, R. (2002). *Intellectual Character.* San Francisco: Jossey-Bass.

Samball, K., Gibson, M., and Montgomery, C. (2007). *Rethinking Feedback: An Assessment for Learning Perspective.* AfL Red Guide Paper 34, Centre for Excellence in Teaching and Learning. Newcastle, UK: Centre for Excellence in Teaching and Learning, Northumbria University.

Shulman, L.S. (2005). Signature Pedagogies in the Professions. *Daedalus*, 134 (3): 52–59.

Vygotsky, L.S. (1978). *Mind in Society. The Development of Higher Psychological Processes.* N.I.Cole, V. John-Steiner, S. Scribner, and E. Souberman, eds. and trans, Cambridge, MA: Harvard University Press.

Wiggins, G. and McTighe, J. (1998). *Understanding by Design.* Upper Saddle River, NJ: Merrill Prentice Hall.

PART III
STRUCTURES THAT SUPPORT INTEGRATIVE LEARNING

7 Linking Integrated Middle-School Science with Literacy in Australian Teacher Education

David R. Geelan

"I'm sorry," she sniffed, wiping her eyes and nose with a crumpled tissue. "It's just that I really have no science background, and your presentation this morning made me really scared. I'm not sure I can pass this course, and I'm pretty sure I can't teach science." Perhaps Shelley's story is not typical of all students in Middle Years of Schooling Science Education, the course I teach to prepare teachers to teach grades 4–9, but I will usually talk to at least one student like Shelley each semester, and know that there are perhaps three or four others in the class who feel the same way but don't approach me about it.[1] The majority of the class will have more science background and be more confident— perhaps too confident in some cases—and four or five will have science degrees and be very confident about teaching science. A large part of the challenge in teaching this course, though, is that many of the beginning teachers I teach will have to teach science at the upper elementary and junior high level soon, and many have little science content knowledge and even less confidence in their ability to understand and teach science.

The Middle Years of Schooling Program (MYS), in teacher education, at the University of Queensland's Ipswich campus (UQ Ipswich) is one of Australia's leading middle school teacher education programs. Modelled on middle school reforms in the United States, but designed from the ground up in the early 2000s for the Queensland school syllabus and context, it is an innovative program that is achieving excellent results. Graduates are sought-after teachers in public and private schools with middle school programs,[2] as well as in more traditional elementary and secondary schools. Middle school teachers are trained as generalists who are qualified and credentialed to teach in all school subject areas.

Participating beginning teachers come from two streams, which are sometimes taught together and sometimes separately. One stream is a "dual degree" program comprising a bachelor's degree in another field such as arts or behavioral sciences combined with a bachelor of education degree. Students spend their first two years predominantly studying in their other field and their second two years studying education. The other stream is a one-year postgraduate diploma in education for those who already have a three-year degree in another field, including science.

This chapter focuses on a core curricular subject from the key learning area of science. It is considered a "curriculum and instruction" course intended to prepare the beginning teachers in the program to teach science to students in the middle years. The course emphasizes developing students' "pedagogical content knowledge" (Shulman, 1986) in school science, including knowledge of the Queensland science syllabus.[3] All students must successfully complete this course in order to graduate from the program. Those who go on to teach in middle school programs will usually teach science, as will those who teach in elementary schools. Some of those who will teach in secondary schools will teach in other specialist subject areas and avoid teaching science.

The beginning teachers' own confidence and knowledge of science are, however, not the only challenges they face in preparing to teach in Queensland schools. The Queensland science syllabus for kindergarten to grade 10 is very open-ended. It prescribes broad general topics that must be addressed, but does not tightly prescribe the order in which they should appear, nor tie specific topics to specific years of schooling. It is an outcomes-based approach to creating a syllabus, and is intended to provide overall guidance for teachers to develop their own rich curricula within school contexts and in response to student needs.

As with any syllabus document, there are trade-offs between specificity and prescription on the one hand and openness and flexibility on the other. Perhaps it is too broad a generalization, but I often think of the balance in terms of this formulation: "tightly prescribed syllabuses support poor (or poorly prepared) teachers but tend to stifle excellent teachers, while highly flexible, nonprescriptive syllabuses tend to mean poor teachers teach poorly, while excellent teachers have the flexibility to achieve excellence." Educators have a wide range of positions on where the best balance lies, but it is true that the quality and amount of science teaching in Queensland schools, particularly at the elementary level, is extremely varied. Some students receive excellent, creative, engaging science education with plenty of experiments and experiences and a good understanding of the nature of science; others receive boring, minimal, textbook-and-worksheet-based science classes—or none. Those coming into the teacher education course have learned in schools from across this wide range of approaches, and therefore come with a wide range of backgrounds, experiences, attitudes, and abilities.

One further challenge in Queensland schooling has been a very strong national push in the past few years for the explicit teaching of literacy and numeracy in elementary and middle schools and for external testing of these skills. This push has tended to lead to very large sections of the school week being blocked out for literacy and numeracy teaching and a reduced emphasis on other learning areas including art, music, languages, and science. Many elementary school students receive only one science class per week, and science is often squeezed out of the curriculum altogether for a few weeks by preparation for tests or other activities.

The Australian Academy of Science, concerned by this reduced emphasis on the teaching of science, adopted something of an "if you can't beat 'em, join 'em" approach in developing Primary Connections, an integrated program comprising both rich science curriculum materials and professional development for teachers. Primary Connections is intended to integrate science education with literacy education at the elementary school level, and is quite explicit in exploring "scientific literacy" and "literacies of science," and considering writing in scientific genres and other literacy-related topics. The academy hopes that in this way science will find a way back into the school curriculum, sharing some of the time in the week that would otherwise be exclusively devoted to the (decontextualized) teaching of "literacy."

The Australian Academy of Science has so far developed 10 Primary Connections curricular packages for units of study (each intended to support approximately 10 weeks of classroom teaching with a few science classes per week) from grades 1 to 7, with topics as diverse as "Weather in my world," "Schoolyard safari" and "Spinning in space." These are examples, rather than an exhaustive curriculum, however: teacher professional development aims to equip teachers with the knowledge, skills, and confidence to develop their own units of study to complement the Primary Connections units. Professional development also focuses on developing teachers' science pedagogy skills, such as asking appropriate questions and guiding student inquiry.

Clearly the set of constraints discussed above on the science education development of beginning middle school teachers and on science teaching in the schools in which they will teach is less than ideal. The relative shortness of the course I teach (discussed below) also offers a less than ideal opportunity to address deficiencies in beginning teachers' knowledge, skills, and attitudes, particularly in relation to the very limited science content knowledge possessed by some. A move is underway in Queensland, in Australia more broadly, and in our teacher education programs at the University of Queensland to reemphasize science education and mitigate many of these constraints, but it will take some years to take full effect. In the meantime, we have to work within the situation and the constraints that apply, and try to offer the best possible preparation to beginning teachers for successfully teaching science.

In attempting to develop and teach a science education course for beginning middle years teachers against the background of these various contextual features and constraints, I drew on my own experience in science education. My background in teacher education before coming to the University of Queensland in mid-2006 was at the University of Alberta, Canada, where for five years I had taught a science education course for physical sciences (chemistry and physics) students as part of a dual degree program. All those students had very strong science backgrounds (the equivalent of a completed bachelor of science degree), enjoyed science, and planned to be senior high school chemistry and physics teachers. They were very confident in their own ability to understand and explain scientific ideas, and excited by the challenge of helping students to understand. The course was very rewarding and I learned a lot, but the challenges were quite different from those I face now in the MYS program.

There are four specific things from my Canadian experience that have informed my teaching in the MYS program at Ipswich:

1. Iteration and continuous improvement: I taught a course with the same course number and the same name for five years in succession in Alberta, but it was never the same course. Each time I attempted to incorporate what I had learned in each of the previous iterations of the course, changing experiences, content, assessment, and other features of the course to better fit student needs and interests and the demands of the profession. During the course I also adapted the activities to follow emergent themes or particular areas of student interest.

2. Three themes: These emerged out of my teaching, reading and reflection and now inform all of my teaching. (1) We teach out of who we are—learning to teach is coming to be. (2) Relationship is the key to teaching. (3) Learners construct new knowledge out of their experiences, on the foundation of their existing knowledge.

3. STSE: An approach to science education that focuses on integrating science learning with studies of science, technology, society, and the environment.

4. Curricular emphases:[4] A particular approach, developed by Doug Roberts and elaborated by Frank Jenkins, to teacher curriculum development and unit planning incorporating STSE.

Education 6550 is a one-semester course for beginning middle school teachers. It consists of nine weeks of on-campus contact (a two-hour lecture-discussion session with the whole class and a one-hour tutorial in smaller groups). Assessment in the course includes planning a single lesson or short lesson sequence revolving around a lab activity and planning a science unit of work (i.e., a science teaching plan for eight to ten weeks of school science classes) or an integrated unit of work in which science plays a large role. At the conclusion of the course all students participate in seven-week, full-time practice teaching sessions in local schools. In the three years in which I have taught the course so far the class size has been 25, 80, and 52 students—these large fluctuations have been caused by changes to the degree programs in which the students are enrolled and to changes occurring at the Ipswich campus. The goals and commitments embodied in my teaching have been the same in each instance, but the practical details of the pedagogy and the students' experiences were significantly different in each iteration of the course because of issues of scale. In the first and third years I took all of the tutorial groups, while in the second year a graduate student facilitated half of the tutorial sessions.

The students have a wide range of knowledge, skills, and attitudes toward science and science teaching when they come into the course, and the goal of preparing them to be able to competently teach science within just nine weeks sometimes seems impossible to achieve. It is fair to say that for some students that is true, and that those students will typically seek jobs in more traditional secondary schools where they will be able to teach within their strongest subject areas (and avoid teaching science). All teachers who graduate from the UQ Ipswich MYS program, however, are considered to be capable of teaching middle school science if their teaching assignment requires it. The task, therefore, is to provide knowledge and skills as fully as possible, but also to work on developing the beginning teachers' attitudes, particularly their confidence and self-efficacy in teaching science.

It is impossible in a course of this length to teach the participants any real quantity of science content knowledge and still meet the goals of the course. We discuss as fully as we can the branches of science and the Big Ideas that underlie them—for example, the notions in chemistry that everything is made out of chemical substances, so chemistry is about all of life, not just colored liquids in glass bottles in sterile labs, and that the features and properties of substances at the submicroscopic atomic level explain their properties at the macroscopic level of human experience. Beginning teachers are also introduced to various avenues for developing their own science content knowledge, including textbooks, other books, popular science publications and Internet searches, and other media, and to approaches to judging the quality of scientific information. Beyond these basics, the beginning teachers are really thrown back on their background knowledge of science, their colleagues' help, and their own willingness to learn more science before and beside their students.

The course, then, is much more specifically focused on helping students develop Lee Shulman's (1986) "pedagogical content knowledge"—not science content knowledge, not generic knowledge about students and teaching (which is delivered in other courses in the program), but specific knowledge about what it means to teach science and how to go about doing it. This includes knowledge of the government-mandated syllabus requirements for science learning, skills in assessing students' understanding of science concepts as well as their skill in conducting experiments and writing scientific reports, knowledge of the nature of science and how to link scientific knowledge with students' life experiences, knowledge of how to conduct laboratory sessions and demonstrations, and knowledge of the ways in which empirical evidence from such experiences can be used to reinforce or challenge students' correct or incorrect scientific conceptions. Students study some of the literature on constructivist theories of knowing and learning and of inquiry approaches to teaching, and learn about planning for instruction in ways that draw on a rich, varied repertoire of teaching strategies and learning activities.

An additional assessment piece—something that was completed for grades, but which was included for the educational value of participating in the activity itself, with grades serving as a motivator—was the development by students of an online personal and professional profile. They developed these profiles using the "Keep Toolkit" (http://www.cfkeep.org), a suite of tools that allows nonprogrammers to develop Web-based profiles of teaching innovations and other things, developed by the Carnegie Foundation. They completed the assignment in three phases, the first two prior to the student teaching block and the third after it.

The first phase involved students answering the question "Who am I?" Responses varied broadly depending on students' self-images and notions of what it meant to try to describe themselves, and included recordings of songs, pieces of video, and photos. One student talked about his childhood in a rough ghetto in Ireland and the journey that took him from there to study philosophy at the Sorbonne and now education in Ipswich. Others talked about pets or siblings, parents with disabilities, karate black belts, religious commitments, and social beliefs. Quite aside from the value for the students of thinking about the different resources that their personality and biography allow them to bring to the teaching profession, this activity was eye-opening for me as the professor, as one more reminder that our students are always more diverse, amazing, and interesting than we could imagine.

This phase, taking place early in the course, meant that it was possible for me to more directly tie in the content and the examples chosen with the students' interests and backgrounds, to draw on particular students' knowledge and experience from their hobbies or earlier careers when asking questions or asking them to present ideas to the class, and to aid them in the process of developing new identities as teachers in addition to their other identities. In tying their learning in the course in with their lives, I was modeling for them in my own teaching the kinds of integrative strategies that I hoped they would develop in their own teaching, and building consistency and coherence between the goals, learning activities, pedagogy, and assessment in the course.

The second phase focused students on the question "Who am I as a teacher?" with subquestions about why they wanted to become a teacher. Again, responses were enormously varied, with common themes including parents or relatives who were teachers, or very inspirational teachers in the students' own educational experience. Another theme, particularly among mature-age students, was that of service to the community and the next generation—one student had been a merchant banker and another an engineer, but they believed that as teachers they would be able to contribute in ways that their earlier careers had not permitted. This sometimes aligned with students' experiences of seeing their own children go through school, and being inspired—either by a great teacher or because they thought they could do better! Almost none of the students responded that they wanted to be teachers because of the longer vacations (a sadly common response from some practicing teachers). The theme of having a vision to serve came through strongly.

Students completed the final phase after they had been practice teaching in schools for seven weeks; it focused on the question "What have I learned about who I am as a teacher?" Many commented on the workload, and the emotional energy required to teach, and many also focused on their own performance and their perceived strengths and weaknesses as teachers. This focus on personal teaching performance is typical of beginning teachers (Fuller, 1969).Others focused on particular facets of the profession that they found enjoyable and compelling, or on others they did not enjoy, and talked about how to develop their careers in ways that would maximize the parts of teaching that appealed to them. Comments were typically at a high level of sophistication in terms of applying a "teacherly" eye to the influences on the classroom, and often indicated a new respect for the profession and appreciation of its complexity. Students frequently commented on their developing confidence in teaching science.

This assignment has been removed from the forthcoming version of the course and replaced by an assignment focused more tightly on the international science education literature and on current issues in science education. It was frustrating to have to remove such a rich integrative learning activity from the course, particularly one that also helped me better understand my students and their goals and aspirations. The choice to do so was in part forced by a reassertion of the very kinds of disciplinary boundaries that the course is intended to dissolve—some students, and some colleagues, had been asking "what does this have to do with science education?" It's true that it has little connection, at least to a narrow conception of science education. It is related more to teaching as an occupation and a vocation and to teacher identity development, and some colleagues argued that the students have other courses, focused on professional issues and pedagogy, more directly focused on those concerns. This specific course is intended to be about pedagogical content knowledge in the field of science education, and it was difficult to justify requiring a quite large time commitment from students that did not seem to directly serve that interest. The course also claimed to help students place their science education practice within the relevant literature, but there was not a piece of assessment in the course that measured that goal. For both of these reasons I believed I needed to remove the online profile assignment—but I will definitely miss it, and am currently trying to develop an in-class activity that meets some of these same integrative goals.

I would argue, however, that through imagining and reimagining themselves as teachers, within the context of a supportive science education course, these beginning teachers were developing (in some cases from a set of base assumptions that would have precluded their imagining such a thing) a conception of themselves as science teachers or teachers of science.[5]

Other activities in the course included completing a number of simple science experiments across the science disciplines. These occurred during the tutorial sessions, and were conducted with a dual focus on having the student teachers do the experiment themselves in the role of students and experience the wonder of discovery and the pain of confusion and then reflect on the process of "teaching in the laboratory." I emphasized that a scientific experiment really only has one "question," and that students should have been prepared in such a way that they have a clear understanding of what that question is, and what specific experimental results will "mean" conceptually. In an experiment testing whether common household materials are acidic or basic, for example, and using phenolphthalien as the indicator, students needed to understand what a pink or clear solution meant in terms of acidity. To understand that point, they also needed to develop a slightly more sophisticated (appropriate to their age and grade) understanding of what an acid is and does.

We conducted experiments, as far as possible, using household materials. Pendulums used nuts and bolts from the hardware store as bobs, rather than the turned brass ones from the university lab. Students conducted chemical tests with small amounts of chemicals and a lot of solutions from home, in ice-cube trays rather than in beakers or test tubes. These choices were very intentional, and resulted from two related concerns. The first was that connecting science with students' own lives and experiences is widely acknowledged as an important part of engaging students with science and making it meaningful for them. The precision offered by "real" scientific apparatus is valuable in science, and in higher level science learning, but much of the science conducted in the middle school is qualitative, or else quantitative at a quite simple level, and using everyday materials such as meter rulers and stopwatches (including the stopwatches in students' cell phones or wristwatches) connected measurements better with students' lifeworlds. The second reason was that many elementary and middle schools are not well equipped with science apparatus. Many do not have dedicated science labs or even science storerooms, and apparatus is often extremely hard to come by. An important part of our concern with this course was to show beginning teachers that this lack of equipment shouldn't preclude their delivering high-quality interactive science experiences for their students.

To link literacy with science, the Primary Connections materials used in the course make a distinction between scientific literacy (Laugksch, 2000, offers a valuable critical review) and literacies of science (Prain, 2006). The materials define scientific literacy as the scientific knowledge, skills, and attitudes that students need to develop in their school science classes and throughout their lives in order to participate as scientifically informed citizens (Geelan, 2010). Literacies of science are the skills of writing, reading, thinking, and reporting that form an important part of science. Prain (2006) speaks of writing to learn in science and describes the ways in which science students use a variety of genres and skills in developing scientific concepts, planning and conducting experiments, reporting their results, and communicating their developing conceptual understandings. The Primary Connections materials essentially use science learning experiences as a context in which to support school students' development of these literacies of science.

This goes far beyond the writing of the traditional, formulaic "lab reports" that might be familiar from past science classes, and in addition to writing includes activities related to reading and comprehending science-related texts in both scientific and popular journals and judging the quality of scientific information. This is particularly important in an environment where memorization of information, while still of some value, is less important because of easy access to electronic reference sources including the Web, but where the information found is of highly variable quality and value. Students in EDUC6550 spent time looking at websites and learned criteria for judging the credibility of sites and information. They also studied scientific claims from media reports, and examples of fraudulent claims. The tutorial groups discussed claims made in relation to controversial issues such as climate change, the health effects of vaccines and living near power lines, genetically modified organisms being used in food, and a wide variety of other issues that the students identified. They explored the links between scientific evidence and social decision making, and studied ways to deal with controversial issues in class without indoctrinating students with teachers' beliefs and opinions.

The literacies of science also include ideas from the philosophy of science about the nature of scientific theories and evidence, the ways in which theories are tested by experiments, and the provisional nature of scientific knowledge. Students compared scientific knowledge and methods with knowledge in other fields of human endeavor, and came to understand the similarities and differences, and the value of all these kinds of knowledge for enhancing life. They explored the features of excellent science teaching explanations, and contrasted these with explanations in science (e.g., in scientific journals) and explanations in everyday life. They explored working with school students to facilitate the students' construction and communication of explanations of scientific concepts and natural phenomena, and learned to assess the sophistication of such explanations.

I often use a fractal metaphor to think about my teaching within teacher education. Fractals are "self-similar across scale"—that is, the fractal as a whole is made up of a number of smaller copies of itself, each of those smaller copies is made up of smaller copies, and so on. In the same way, my teaching of my students in my course is mirrored by their teaching of their students in their courses. I in turn learn both from my own teachers—colleagues and those whose work I read—and from my students, and they learn from their students. This means that my own pedagogy in the course should mirror the kinds of pedagogy I hope to have them learn and apply in their own teaching (with a few differences because my students are adults rather than adolescents). There should be consistency between the goals of their learning in the course and the goals of the students they teach, and my students' pedagogy should be consistent with those goals. Further, their teaching activities and development of teacher identities constitutes a progress of integrating what they learn in this course with their existing identities as people, friends, siblings, spouses, and so on, the things they learn in their other university courses and practice teaching sessions, and the things they learn in their daily lives and that they learned in their schools or jobs before coming to the university. Their own pedagogy should focus on developing these same kinds of rich integrations for their students.

"Shelley" never really came to love the idea of teaching science—her final profile posting doesn't even mention the word. It does say that she feels ready and equipped to be a teacher, and outlines her commitment to helping students develop their critical reasoning. But she confided in a conversation toward the end of the semester that she had conquered her fear of science and science teaching. Although she would be seeking a job as a junior high art teacher and was unlikely to teach science, she did believe that she would be able to do so if asked, and also said she thought herself much more capable of reading and understanding science reports in the media, and much less intimidated by science and scientists. Other students, like Cassandra, had moved even further—from a position where they had essentially no exposure to science to one where they had been given the opportunity to teach science while practice teaching, had worked hard and prepared, and had been successful in teaching new science concepts to their students and in integrating science with literacy and with students' life experience.

Notes

1. All names used are pseudonyms.

2. Approximately 10% of Queensland schools have designated and structured middle school programs, with a government-mandated push to move toward more middle schools currently in progress.

3. I use "syllabus" to refer to the state-mandated list of content to be taught and "curriculum" to refer to the school- and teacher-developed program of learning activities for students.

4. Geelan (2010) discusses the curricular emphases approach further, with examples.

5. I make this distinction because as middle school generalists these graduates typically think of themselves as middle school teachers first, and that identity then encompasses being teachers of a variety of more or less integrated subject areas, including science.

References

Fuller, F. (1969). Concerns of Teachers: A Developmental Conceptualization. *American Educational Research Journal*, 6 (2): 207–226.

Geelan, D.R. (2010). Science, Technology, and Understanding: Teaching the Teachers of Citizens of the Future. In R. Nowacek, J. Bernstein, and M. Smith (eds.), *Citizenship across the Curriculum*. Bloomington: Indiana University Press.

Laugksch, R.C. (2000). Scientific Literacy: A Conceptual Overview. *Science Education*, 84 (1): 71–94.

Prain, V. (2006). Learning from Writing in Secondary Science: Some Theoretical and Practical Implications. *International Journal of Science Education*, 28 (2–3): 179–201.

Shulman, L.S. (1986). Those Who Understand: Knowledge Growth in Teaching. *Educational Researcher*, 15 (2): 4–14.

8 SCALE-UP in a Large Introductory Biology Course

Robert Brooker, David Matthes, Robin Wright,
Deena Wassenberg, Susan Wick, and Brett Couch

WHEN THE FACULTY of the College of Biological Sciences at the University of Minnesota reviewed their introductory biology curriculum in 2003–2004, they found courses that were being taught in much the same way as they had been for decades. The courses were thoughtfully, conscientiously, and enthusiastically taught. However, the innovative approaches to teaching and learning that had emerged in the preceding decades, and the increased understanding of the process of learning that provided the foundation for those approaches, hadn't made their way into the large lecture courses dedicated to introducing students to biology and what it means to be a biologist and do biology.

A key shortcoming was skill development. Beyond learning the basic knowledge in the discipline, students also needed to be trained in the skills necessary to succeed as a biologist. These include using online bibliographic and bioinformatic databases to explore primary literature and molecular data, designing authentic biological experiments, writing scientific papers, and working effectively with members of a collaborative team. Many faculty members thought that the drive to achieve economies of scale by adopting large class sizes had prevented them from teaching their students the skills required of working biologists. In short, the faculty wanted to transform the learning setting and pedagogy so that students engaged in the authentic practices of biologists in the real world. This aspiration, which in this case is highly disciplinary, is part of the connected science vision to have students engage in course learning as scientists do in research teams.

By 2006, the faculty had decided to overhaul the introductory biology course for majors, developing a new approach that included substantial attention to skill building. While the core biological concepts for the course remained similar, very little else would be the same. In fall 2007, biological sciences majors enrolled in the first offering of this transformed course, called "Foundations of Biology." It is a two-semester course, consisting of Biology 2002 and 2003/2004 (Biol2002 and Biol2003/2004). Biol2002 includes a laboratory component. Biol2003 does not include a laboratory, but the students also enroll in Biol2004, which is a research-based laboratory that students must take concurrently with Biol2003. The laboratory components of Biol2002 or Biol2004 also emphasized a combination of skill development, hands-on experience with central course concepts, and the pursuit of authentic research questions. In this chapter, we focus exclusively on the "classroom" component of the course.

Our Foundations course is required of all biological sciences majors. Each semester, we offer two sections of Biol2002 and 2003, each with 100–130 students. The course meets in 115-minute sessions, three times a week for Biol2002 and twice a week for Biol2003. Both semesters are taught by a pair of instructors who are committed to student-centered learning, use a team-based learning approach (Michaelsen et al., 1982), and meet in a SCALE-UP classroom (Beichner, 2000). However, the two semesters were designed in part by different instructors and cover diverse content. As a result, some details differ. These differences, we believe, illustrate real strength in the variety possible when using a SCALE-UP classroom for biology and science learning and teaching.

In this chapter, we describe how our introductory course has adopted the SCALE-UP model (Beichner, 2000), incorporating principles of student intellectual development (Bloom et al., 1956) and team-based learning (Michaelsen et al., 1982). We then discuss the types of specific learning activities—what we will call tangibles (hands-on) and ponderables (minds-on). Given that an integral part of the SCALE-UP model is the intensive use of technology, we discuss our integration of technology, followed by a discussion of student assessment, particularly the formative assessment and the assessment that promotes team coherence and effectiveness. Our implementation of SCALE-UP is evolving and programmatic assessment is in the early stages. Thus, we conclude with a reflective discussion of our experience and an exploration of issues and challenges.

Authentic Biology: Team-Based Learning

Inspired by the idea of highly effective student teams for student learning even in large classes, we adapted Larry Michaelsen's team-based learning instructional strategy (Michaelsen et al., 1982) to biology in a SCALE-UP classroom environment. This kind of powerful learning is characterized by working in teams with people of varied backgrounds who learn to appreciate the value of different perspectives when solving complex problems, a hallmark of connected science. We create student teams that are as diverse as possible in terms of gender, country of origin, and prior biology experience. We aim for each team, however, to have about the same heterogeneous composition.

Rather than implementing Michaelsen's optimal team size (five to seven students), we follow the pattern developed by Beichner's original SCALE-UP class design. As a result, our students are typically placed into teams of nine. Students then naturally form subgroups of three students, since the table consists of three arcs, each with three chairs and a laptop computer.. In addition, the nature of the project work often leads students to form subgroups with a "divide and conquer" approach. Even so, interactions during the class, including problem solving, data analysis, and the completion of projects involve interaction among all students in a team. Students remain in the same team for the entire semester. Permanent teams are integral to our Foundations courses because some projects last the entire semester, and because students need time to learn to work effectively in teams and trust their fellow team members. Because of the developmental stage of the students with regard to teamwork skills, they also need time to learn about giving and receiving constructive criticism. Of the approximately 360 teams we have set up so far, only a few cases have arisen in which team members have been reassigned because of conflicts within a group.

How do we choose the teams? In standard team-based learning, this occurs on the first day of class by grouping students based on relevant or random factors such as prior subject knowledge or distance of student birthplace from the university. While these strategies might be easier and more transparent to the students, in Biol2002, we typically assign teams based on the results of an online survey that students take prior to the term's start. The survey includes a question asking each student to choose, from a long list of adjectives, a single adjective that best describes him or her. We then use this information, in combination with the other factors mentioned above, to establish teams that also have a diversity of personal self descriptors including "hard-working," "creative," "leader," and "friendly" (Wright and Boggs, 2002). We find that the diversity within each team helps students learn that people who look, sound, and engage the world in very different ways have valuable contributions for the common good. For example, on course evaluations and reflections, students mention the deeper global perspectives they gained from their peers who are from another country.

Others ways in which we have applied team-based learning principles to the SCALE-UP model are by using individual and team readiness assurance tests that we call learning readiness quizzes (LRQs) at the start of each new unit, having students engage with significant activities (working on the same questions or problem and then reporting their answers to the entire class) rather than lecturing to cover material, and having team members evaluate the contributions and effectiveness of teammates, the outcome of which makes a significant contribution to student individual grades.

Authentic Biology: Promoting Higher-Level Understanding

Both the team-based learning and SCALE-UP classroom approaches recognize the importance of moving the learning of basic knowledge, at the lowest two levels of Benjamin Bloom's cognitive scale (Bloom et al., 1956), to outside of class. While basic knowledge is important, we accept the research findings on the limitations of lectures for conveying this information (Handelsman et al., 2006). Instead, we expect our students to learn the basic concepts from the text before the start of each course module. Class time then emphasizes application and analysis that promotes higher-level understanding. In both Biol2002 and Biol2003, traditional lecturing is typically less than one hour per week and is often conducted in short blocks interspersed with learning activities.

In the course, instructors engage students in distinct types of learning activities: "tangibles" (hands-on) and "ponderables" (minds-on). Biological tangibles can be divided into those most suited for a "wet" technology laboratory setting and those suited for the SCALE-UP classroom's "dry" technology environment. Biol2002 and Biol2003 have separate wet lab sections called "research lab" that allow students to do lab-based biology. For example, students use microscopes, carry out electrophoresis, and culture bacteria. In what normally would be called "lecture," students engage in a "concept lab," a term borrowed from Daniel Udovic's Workshop Biology (Udovic, 1996). Here, students work actively with data, simulations, case studies, and physical models. Ponderables—pencil-and-paper problems, discussion questions, and case studies, for instance—are also well suited for the SCALE-UP classroom where thinking like a biologist and in a team defines the learning.

In both semesters of our course, students also work on projects that span from four weeks to the entire semester. These projects provide students with the experience of doing the sustained intellectual work of a biologist. Students engage the primary literature, work collaboratively in a team at every stage, use evidence-based, persuasive scientific writing, and present their final work to peers at a poster session.

Although evaluation of projects requires carefully constructed rubrics that take significant time to develop, projects are very effective in getting students to think like scientists, to grapple with biological concepts at the highest of Bloom's cognitive levels, and to integrate concepts from all sections of the course. For example, in the first term (Biol2002), students are charged with writing a proposal to create a genetically modified organism that would offer some social benefit. We make the semester-long project manageable by breaking it into stages on which we can give context-specific instruction, formative feedback, and a chance for intrateam and interteam peer review. For the cell biology section of the Biol2003 course, the long-term project is a mini–grant proposal in which each team proposes two novel hypotheses regarding a transport protein, and designs experiments to test those hypotheses. Each team turns in a 15-page paper, excluding figures and references, organized like an actual grant proposal.

Authentic Biology: Technology

A critical feature of the SCALE-UP course design is that it takes place in a technology-enabled classroom and uses that technology to promote communication, collaboration, and the exploration of ideas. A key aspect of many projects is the ability of students to use their laptops in class to identify appropriate information sources such as journals and government websites. This enables them to expand their learning beyond the bounds of a traditional textbook. We also coach them on effective search strategies and help them develop information literacy, including the appropriate use of Wikipedia and other online sources.

The classrooms have the basic "banquet hall" design of the SCALE-UP model, with 13–14 round tables, each seating nine students. Each table has two or three microphones and connection ports for three computers (see Figure 8.1). We provide the class with three laptops per table, though many students bring their own laptops. Depending on the activity, we may limit students to use of the classroom computer only, minimizing the temptation to use their personal computer for purposes outside the scope of the class activity. Each table has a large dedicated LCD display that students can use to work on a problem or task with their team. The displays also allow team answers or work to be displayed simultaneously to the entire class, allowing whole-class discussion, productive interteam rivalry, team accountability, and sharing of examples from a single team with the entire class. Each team also has a whiteboard area for brainstorming, concept mapping, and outlining their work. It is remarkable how focused students become when members of their team are working at their whiteboard.

Overall, instructors for the Foundations course have observed a change in the team dynamic when students are using these technologies ("Inside an Active Learning Classroom," 2012). Everyone seems more engaged. The team has a focal point—a concrete task—for their efforts. For example, the entire team will work together to edit their papers or improve their poster.

Students use the laptop computers for simulations of biological processes, designing plasmids, bioinformatic work, collaborative writing, and exploring the primary literature. They use the laptops extensively on their projects, gathering sources and data, forming bibliographies, writing and editing text, and putting together a poster for presentation to their classmates.

Websites play a critical role in organizing and facilitating student activities. Each course uses a WebVista or Moodle site where students find all class materials posted, including study guides, self-tests, class presentations, activities, grading rubrics, website animations, and extra readings. Students turn in assignments by uploading them to a course management system drop box, take self-tests to check their readiness for the weekly learning readiness quiz, and check their grades over the course of the semester. We also use SurveyMonkey to administer surveys at the beginning and end of the semester to gather information on class readiness and experience.

Our students often live far from each other and have very full schedules that may preclude them from meeting as a team to work on their projects outside class. As a result, teams typically use collaborative online tools such as Google Docs or the group networking tool Wiggio to facilitate student collaboration. Students post projects that are in progress so that other students may evaluate them. In addition to these student-managed sites, we have set up team blogs at our course management site where students can exchange information and documents. We have also recently begun experimenting with virtual conferencing capabilities.

Finally, the technology of the instructor's podium is crucial for the facilitation of class activities. The instructor can present slides from a laptop, show videos with sound ported to classroom speakers, manage the screens and displays, and display work created in real time with a document projector that can send images to all room displays. The document projector also comes into play when teams work at their tables with pencil and paper. For example, when a team comes up with a concept map or outcomes of Mendelian crosses, it can display these documents to the class via the projector and discuss the work with many others.

Authentic Biology: Assessment

Assessment is a critical component of our course. Consistent with our goal of teaching authentic biology, our assessments mirror how feedback, monitoring, and review are incorporated into the natural and staged processes of doing biology as a research scientist. Formative assessment is woven into the assignments, and it comes in the form of self-assessment, peer review, and instructor feedback. We consider an essential part of a good course to be opportunities for students to receive frequent feedback on their work, and to have chances to improve not only their work but also their grade by showing they have learned from the feedback they received. Summative assessment also is an important component that comes in the form of take-home exams, in-class exams, and papers.

In the first semester course, learning readiness quizzes are given at the beginning of every week. According to Michaelsen's team-based learning strategy, we follow that quiz immediately with the same quiz taken as a team with instant feedback on team answers given by Immediate Feedback Assessment Technique (IF-AT) scratch-off style answer forms (Epstein Education Enterprises, Cincinnati, OH). We follow these back-to-back quizzes with discussion of remaining misunderstandings, setting the stage for team activities at the heart of the SCALE-UP and team-based learning approaches. While Michaelsen's team-based learning strategy recommends six or fewer readiness assurance tests in a semester, we have increased the frequency to allow reading assignments to be briefer (often one chapter) and the mastery of course material (course objectives) to be assessed in greater depth. At the end of the semester, students are also able to take a cumulative quiz of 100 questions that are representative of the questions from the semester's LRQs. If students do better on the cumulative quiz than on the sum of their individual LRQs, they can substitute that better score.

The second semester course takes a slightly different approach. We ask students to read material ahead of time and to ponder "weekly questions" that take a deeper look at the textbook material or include a more experimental focus. Following a class discussion of the material, we give a quiz later in the week that incorporates both the basic textbook and material discussed in class.

Both semesters of our course have three or four summative assessments (i.e., exams), on which application and analysis questions predominate. Depending on the instructor, these exams can be take-home, partly take-home, or in class. Those of us who use take-home exams are impressed with the amount of learning that can occur when students engage with challenging questions over a period of a week or so. Once an exam is completed, students can earn back a quarter of the missed points by completing a "postexam analysis." For each question in which they did not receive full credit, they must analyze in writing how their answer is different from the exam key, how it would need to be changed to receive full credit, andhow they can use this information ation to improve their performance on a similar question in the future. They also complete a global analysis of their performance on the exam, explaining the best qualities of their answers, how they could improve their overall performance, and the new strategies they will use to prepare for the next exam.

Course activities, such as the writing projects, also receive a great deal of feedback from both peers and instructors. Some assignments are broken down into sequentially produced subproducts. For example, this might include three ranked and justified topic ideas, a bibliography, a background section, a research plan, a summary section, and a poster. Peer evaluation occurs by subgroups within teams and among teams. Instructors give feedback, including detailed scoring with a rubric and written comments to help teams improve their scientific writing and, more importantly, their scientific thinking. One of us provides feedback via screencasts created with software such as Camtasia or Breeze. Students value this visual and verbal feedback and the faculty member finds that she can provide more detailed and nuanced feedback, particularly encouragement about what they did well.

In the second semester, formative assessment of projects also takes place via a repetitive peer-review process. At the beginning of the semester, each student signs up to be responsible for turning in three or four team projects on particular dates. Students primarily discuss these projects on Tuesday. On Thursday, the student responsible for turning in the project brings a written draft to class, and the entire team revises and refines the project. The team then submits the project online on Friday.

In both semesters, peer evaluation of contributions and effort affects a student's grade; approximately 6% of the final grade depends on this evaluation by a student's team members. As a result, a great incentive exists for team members to work hard to contribute meaningfully to their team project and other activities. In keeping with the value placed on giving formative assessment, a low-stakes peer assessment in which the grade does not count is given at least once in the semester, often in the fourth or fifth week. This feedback serves as a formative assessment for their teamwork skills and we coach students on using it to improve their performance for the remainder of the semester.

In both semesters, team points account for about 40% of the student's course grade. However, these points usually do not vary much from team to team. As a result, as in traditional courses, most of the variation in student grades stems from differences in individual performance. Working within a team surely motivates individuals to do their best work, in part because the team is depending on individual contributions and will be evaluating team members based on them. As a result, most team scores cluster in the high end of the range (B+ to A quality work).

Though most of the variation in student performance can be traced to variation in individual performance on quizzes and exams, we have evaluated scoring in the second-semester course. We see a trend in which lower-scoring students tend to get pulled up most by their team scores. We have some concerns that this observation reflects grade inflation for the weaker students. However, it is also possible that certain students with poor test-taking skills end up with a grade that more fairly represents their understanding.

Opportunities and Challenges

Though our introductory biology course draws on a diverse set of pedagogical inspirations, the contribution of the SCALE-UP classroom model, and the pedagogies that it naturally promotes, is very pronounced. The room design brings the student focus away from the instructor and to the teammates. Having students sit at round tables is a surprisingly profound change. Face-to-face discussion among students becomes a primary vehicle of instruction and learning. Also, the technology promotes interactions in different ways. Team discussions may focus students' attentions on a whiteboard, an individual's laptop, a large-screen display, or an online collaboration site. In addition, the available technology permits whole-class discussions, which the instructor can easily facilitate.

The SCALE-UP design consciously dismantles the tidy instructor-oriented instructional model, with all eyes forward on the instructor and instructor-controlled media. Instead, students are expected to be engaged, working together with classmates on activities, coming to consensus by considering the diverse opinions of their peers, and creating something new with what they've learned. In other words, the pedagogy involves intentional integration of individual learning, learning in a diverse team, and the creation of concrete and authentic products. The student-centered classroom feels much more vibrant than a lecture hall. At the same time, the louder and visually distracting environment may feel somewhat freewheeling to students who are more comfortable with a linear lecture. Some students who are easily distracted have reported that it can be difficult to focus in a busy room. These are inherent trade-offs that come with the SCALE-UP course design and student-centered course style. We believe they are trade-offs worth making.

Student reaction to the courses has been mixed but has improved with each passing semester as we are better able to calibrate workload and preemptively address concerns. Some are thrilled to be in a class where they are held accountable for reading prior to class and where they are engaged in activities that allow them to analyze data and apply what they have learned. They can work collaboratively with other talented students to accomplish things they would be otherwise unable to do on their own as first- and second-year students, such as developing a research proposal or a poster. Others resent having so much expected of them, resist having any part of their grade be based on team assignments, and distrust the exchange of lecture time for everything else that the course offers in its place. This range of responses may also be related to the intellectual development of students. Students farther along in their development tend to be more comfortable and adept at learning that demands more active and higher-order thinking.

Optimizing the functionality of teams is clearly a challenge. In one anonymous survey, 69% of respondents indicated their team worked very well or moderately well together. Even so, we are concerned about the other 31% of students who were less positive about their team function. What is their perception of their teamwork? What can we, as instructors, do to improve the team function?

The careful choice of activities, transparent organization of class time, and frequent explanation of the learning objectives have helped us create an atmosphere that most students appreciate. Instructors need to repeatedly spend some time reminding students why we believe this paradigm shift in pedagogy is worthwhile: because it fosters skill development and higher-level learning. Nevertheless, we accept that some students will not appreciate a class with a nontraditional format that is currently (and unfortunately) unusual for them.

The SCALE-UP biology course has allowed us to introduce our biology students to the real work of being a biologist in their first major college course in the discipline. Our students are acquiring skills that are needed by practicing biologists, in addition to obtaining a deeper understanding of concepts in biology. The SCALE-UP model is based on an understanding that people learn best by doing and by teaching others. The physical classroom and course design both facilitate this kind of learning. We accept that the experience may be more challenging for some students, including students with poor English fluency and students who are particularly poor readers or poor communicators. Yet we recognize that awakening students to the need to improve in these areas early may be very helpful to them as they prepare for upper-division biology courses and their work after graduation. We are aiming to provide students with a transformative experience in biology, one that imparts the foundation for their future coursework and work in the field and as lifelong learners.

One persistent faculty concern is the degree to which content must be reduced to make way for a student-centered approach that includes more active-learning experiences and teamwork. If you have four to six hours of class per week and spend less than one hour conveying knowledge to students, how can you possibly cover the necessary amount of information? With clearly defined objectives, our course faculty believe a course can be designed so that students can interact with course content in diverse ways: in the study guide, in the reading, in online self-tests, in the learning readiness quiz, in activities, in discussions of those activities, during project work, on exams, and in the accompanying research lab. Moreover, the reinforcing exposure to concepts is responsive to the difficulty of a concept. Concepts that are easy to master don't elicit as much discussion among teammates, don't require discussion with the class a whole, and don't need to be included on an exam. Paired with the quality of learning that comes from working on activities, solving problems, and applying knowledge to new situations within their project, our approach helps students encounter the material in multiple venues and makes the "coverage concern" a minor one. In fact, we think we cover the material in a deeper, more nuanced way than we could do in a lecture-only format, and consequently promote greater retention of the material.

Our introductory course is still being refined and (at the time of this writing) has been in operation for only four years. In this chapter, we primarily share our ideas and observations, rather than try to persuade with quantitative data that the SCALE-UP model works for an introductory biology course. Our preliminary results are promising, however, and suggest that students who have completed our new Foundations course earn significantly higher grades in traditional-format upper division courses and are retained in biology at a higher rate. As we mentioned, student ratings of teaching are somewhat mixed. However, this could be because course instructors are purposefully stepping more into the background and letting students explore ideas and converse among themselves in search of understanding. When the instructor becomes a "guide on the side" rather than the more familiar "sage on the stage," we would expect ratings of an instructor to be more mixed, if only because the instructor is lecturing in the limelight less. Students may also have a hard time transitioning from their prior experience of science learning in a lecture format to one that requires them to be much more engaged.

At the University of Minnesota, the SCALE-UP style is beginning to spread to other biology courses. Some sections of introductory genetics and cell biology are now being taught in a SCALE-UP format in active learning classrooms, as are sections of two introductory biology courses that satisfy liberal education requirements for nonmajors: Evolution and Biology of Sex, and Our Global Environment: Science and Solutions.

Furthermore, a new building at the University of Minnesota–Twin Cities campus that houses 10 technology-enhanced classrooms with round tables appropriate for student-centered teaching opened in fall 2010. As more faculty gain experience and see how effective and stimulating the SCALE-UP model can be for student learning that mirrors the real practices of biologists, we anticipate that many will realize its value and change their current lecture-based model of teaching.

References

Beichner, R.J. (2000). Student-Centered Activities for Large-Enrollment University Physics (SCALE-UP). Proceedings of the Sigma Xi Forum "Reshaping Undergraduate Science and Engineering Education: Tools for Better Learning," Minneapolis, MN, Nov. 5–7, 1999, pp. 44–52.

Bloom, B.S., Englehart, M.B., Furst, E.J., Hill, W.H., and Krathwohl, D.R. (1956). Taxonomy of Educational Objectives: The Classification of Educational Goals. *Handbook I: The Cognitive Domain*. New York: Longman.

Handelsman, J., Miller, S. and Pfund, C. (2006). *Scientific Teaching*. New York: Freeman.

Inside an Active Learning Classroom. (2012). College of Biological Sciences, University of Minnesota. Available at http://www.classroom.umn.edu/projects/alc.html. Accessed Nov. 22, 2012.

Michaelsen, L.K., Cragin, J.P., and Fink, L.D. (1982). Team Learning: A Potential Solution to the Problems of Large Classes. *Exchange: The Organizational Behavior Teaching Journal*, 7 (1): 13–22.

Udovic, D. (1996). *Workshop Biology Curriculum Development Handbook*. Eugene: University of Oregon.

Wright, R. and J. Boggs. (2002). Learning Cell Biology as a Team: A Project-Based Approach to Upper-Division Cell Biology. *Cell Biology Education*, 1: 145–153.

9 Reuniting the Arts and Sciences via Interdisciplinary Learning Communities

Xian Liu, Kate Maiolatesi, and Jack Mino

For more than a decade, Holyoke Community College (HCC) has been helping students pursue their learning in more intentional, connected ways using interdisciplinary learning communities (LCs). As a self-described "learner-centered institution," HCC's mission statement identifies LCs and interdisciplinary courses as two of the "contemporary assortment of instructional strategies" supported by the college. Our campus mission of providing access, equity, and excellence in teaching and learning infuses the Learning Community Program's mission—to provide interdisciplinary learning communities to promote integrative learning across disciplines in the general education curriculum and career programs, and affirm the value of community for increasing student involvement in learning.

Learning communities engender competence in both students and faculty, with their explicit valuing of relationship and community, emphasis on collaborative teaching and learning, shared epistemology, and integrative assessment. Crosscutting texts, conceptual organizers, integrative reading, thinking-writing-discussion prompts, seminars on primary source texts, and collaborative-integrative projects are all examples of the instructional strategies LC faculty use to foster interdisciplinary learning. The LC we taught for seven years—Sustainability: Surviving the 21st Century—integrates English 101 (ENG101), a first-year composition course, and Introduction to Sustainability Studies, the first course now in the Sustainability Studies Program. The course description during 2003–2010 goes as follows:

> As one community, the Earth's inhabitants are faced with many critical problems in the 21st century—extinction, diminishing energy resources, increasing population,

and human civilizations' limited vision of alternatives. Whether *Homo sapiens* can learn to make sustainable choices will impact the long-term survival of all the species on this planet. The challenge of finding and forging a sustainable relationship with the Earth animates this Learning Community (LC). The course offers general background on ecology and biodiversity. Participants will explore in class discussion and expository writing three major focuses in the current environmental debate: 1) sustainable agricultural practices, 2) green building practices and 3) sustainable energy systems. Student-led seminars, laboratory experiments, community-based learning activities, and expository writing are all integral components of the course. In its various rhetorical modes, expository essays will be assigned, discussed, and practiced within the context of all learning activities.

In keeping with an integrative approach to science education, our chapter begins with a story—the academic and personal journey of three LC students. We then examine the opportunities and challenges of this interdisciplinary LC model in pedagogy and curriculum design. Finally, we conclude with a personal reflection on LC teaching, learning, and faculty development.

Sustainable Development—Students' Stories

This is a story about the academic and personal journey of three Sustainability LC students, Jane, Carol, and Debbie from the fall of 2005. Jane, a traditional age student, entered the class, uncertain about what she wanted to do in life or what her strengths were. She had yet to declare a major, but knew it would not be in the sciences. Carol was an accounting major. She had a family and wanted to finish school as soon as possible. She had grown up on a farm and always thought of the natural environment in utilitarian terms. She saw nothing wrong in her view and was confident that she could accomplish the dual mission of finishing her English and lab science requirements in this course. Carol saw writing as her strength. Debbie, in her midthirties, was coming back to school after raising a family. She had been doing some work as a naturalist and wanted more education in the field of environmental science. Although she knew she could contribute to the class, she felt a little disconnected from the other students in the beginning.

Soon after a midterm project, these three women decided to work together on their final project. The assignment was to offer suggestions for HCC to reduce its ecological footprint. They chose to focus on four aspects: habitat restoration, green roofs on existing buildings, a water catchment system, and natural ventilation systems in the campus buildings. The course's integrative approach required them to conduct research on their topics as well as effectively articulate their ideas and proposals visually and verbally. The completion of this project consisted of a research paper, a color poster, and a group presentation.

Each student brought her own expertise to this project and helped to make the process easier for all. Carol was the primary writer; Jane produced excellent posters with the use of PhotoShop; and Debbie's natural affinity for the science helped direct the project. The students proposed starting a habitat restoration program and retro-fitting green roof, natural ventilation, and water catchment systems on the campus buildings. Habitat restoration would use reconciliation ecology to conserve species diversity on campus; green roofs alone could save 25% in cooling and heating cost in one growing season; retrofitting solar-powered vents high in the buildings would fundamentally improve the indoor air quality; and water catchment systems would store rainwater runoff with a series of gutters and containment drums to reduce soil erosion on campus. Their project provided not only the vision but also the formula for calculating potential savings for HCC.

The project was first presented in class, with faculty and administrators in atten-dance. The audience was small but engaged, as the project took a realistic look at actual problems on campus. The audience responded thoughtfully. Following the class pre-sentation, the students were invited to present their plan to the Holyoke Community College Sustainability Committee—a state-mandated committee dedicated to reduc-ing the ecological footprint of our campus. The committee was very impressed and made plans to replace invasive plants with native perennials and retrofit buildings with a runoff catchment system. The students were amazed that others were serious about their ideas.

Following their work at HCC, all three women have gone on to other institutions. Jane is now in the bachelor's degree program of environmental science at the Uni-versity of Ohio. Debbie, still passionate about science, is at Smith College on a full scholarship, continuing her pursuit in environmental science. Carol has completed her accounting degree at the University of Massachusetts. Although she has entered a field far removed from environmental science, she cannot help thinking about recy-cling and waste reduction in the business world just as her final "course reflection" says (2007) that she feels so much more connected to the farm she has grown up on because she has a better understanding of what must be done to care for it.

Integrative Pedagogies for Learning Communities

The theme of sustainability has inherent "teaching" value to both introductory science and English composition because it naturally lends itself to a wide variety of science projects and writing assignments. While both instructors, one from English and one from the Department of Environmental Science, are committed to teaching science through writing, the interdisciplinary pedagogy was far from intentional in the begin-ning. However, integrated assignments and assessment strategies have evolved since it became clear that science could offer college composition courses realistic content and ENG101 could lead students to reflecting on science through their writing.

Writing well across disciplines promotes a student's success in college, while acquiring science-based analytical skills encourages the development of critical think-ing. A college science paper is typically most concerned with content. Lab reports are usually written to report findings, and research essays are to gather facts, report trends, and test hypotheses. The quality of the conclusion often determines the grade of an essay. Neglected are the other aspects of effective writing such as brainstorming for topics, constructing arguments, organizing ideas, word choice, and grammatical correctness. On the other hand, writing skills can be more instrumental if they are developed as a means of reflection and communication rather than just a reporting device. Thus, integrated assignments and assessment strategies are intended to use writing for thinking and use scientific principles to back up a writer's claim.

All writing assignments, from low-stakes journal entries and seminar papers to formal essay topics, are designed to promote the understanding of sustainability by bridging science learning and college composition. Since some of the main academic challenges for the first-year students include superficial comprehension of scientific principles, lack of ability to apply science in discussions, and misconception of disciplinary boundaries, not to mention their paranoia over writing, our writing assignments purposefully "ignore" the disciplinary divides between English and science by insisting that students discuss readings and issues in writing. Here is the list of sample assignments we have used:

- Identify a "sustainable practice" you have observed and explain through specific examples the defining characteristics that make that practice "sustainable." (Rhetorical mode: Example)
- Explain the process of converting solar energy into electric energy. You may want to focus on the physics of the conversion if you decide to write on this topic. (Rhetorical mode: Process)
- Economic reasons are often cited as one of the main contributing factors to humanity's insensitivity toward the degradation of the environment. Please explore some of the possible reasons/causes in the culture of an industrialized country that may lead to the nonchalant attitude toward the health of the environment. Be specific in your supporting examples. (Rhetorical mode: Cause/Effect)
- Economic reasons are often cited as one of the main contributing factors to humanity's insensitivity toward the degradation of the environment. Please explore some of the possible reasons/causes in the culture of an industrialized country that may lead to the nonchalant attitude toward the health of the environment. Be specific in your supporting examples. (Rhetorical mode: Cause/Effect)
- Freshwater is a precious resource although renewable on Earth. Compare and contrast the water consumption pattern you have observed in one household with the most efficient ways of using water as discussed in Miller and/or Corbetts. (Rhetorical mode: Comparison/Contrast)

For integrated assessment, we have devised rubrics as "grading criteria" (Table 9.1) to make our expectations for these integrated writing assignments explicit. We ask students to grade several sample essays using the rubrics before they turn in any of their own papers. Through this exercise, they can get a clearer understanding of the grading rubrics, and the instructors a more realistic calibration of student understanding. The rubrics are an effective tool for integrating writing in science because their categories separate what seem to be various fuzzy components of a paper into structural unity, support, content synthesis, sentence skills, and so on. As a result, a grade on an essay no longer offers just a general impression but specifics for acknowledgment and improvement. Each student can now concentrate on those areas where his or her paper is still weak, but students also know the areas where they have done well. The "typical" feedback on the last page often reads: "This paper is strong in Unity and Coherence as it has a clear thesis statement and uses transitional sentences and signal phrases effectively. But the Content & Synthesis needs more work, because it doesn't quite explain the inner workings of the conversion from sunlight into electricity." In other words, neither a paper "big on theory" nor a report data laden without the big picture could score high according to the rubrics. Since most of the formal essay assignments have the option of revision, specific end comments like the above are particularly welcomed by those who want to revise their works.

Community-based learning projects (CBL) form another pedagogy that can help students integrate theory and practice in science and engage with the values of science. We were not able to take the course into the community in the first year, as we were preoccupied with bridging our two disciplines. But as the integrated strategies and techniques became more established, we were able to add more experiential components to the course.

The first CBL project is to learn about sustainability at a local food bank farm. The food bank farm is an organic farm that donates half of its produce each season to the local food bank, which offers this food free to community members who might otherwise go without fresh fruits and vegetables. Each student is required to spend 15 hours working on the farm during the semester. Also, the laboratory portion of the course (2.5 hours per week) is connected with the farm work whenever possible. With the help of the farm manager, we have designed labs on topics such as composting, cover cropping, and seed saving. The class is given readings on the subjects of the farm manager's lecture before the students perform any specific task for the farm in the area concerned. These immediate hands-on experiences seem to have made abstract concepts more accessible to the students.

For example, increased agrodiversity can strengthen the topsoil and create favorable growing conditions, whereas the single-crop arrangement weakens the topsoil and depletes its nutrients. This can seem counterintuitive to students. One crop taxes the soil while diverse plants complement one another and enrich the soil. Textbook examples and classroom discussions can certainly explain the reasons behind the concept. The food bank farm offers an alternative. After a few times working on intercropping on the farm, George wrote in his cause/effect essay "Sustainable Farming": "It is possible for the [food bank farm] not to use as many chemicals as other farms

Table 9.1. Integrated Grading Criteria

Score of 5 indicates that the essay meets all or most of the following requirements with distinction; score of 3 indicates that the essay adequately meets most of the requirements, but needs more development; and score of 1 indicates that the essay does not meet most of the listed requirements.

	5	4	3	2	1	n/a
Unity (focus / main point / overall essay structure / clarity)						

- The introductory paragraph clearly identifies or states the main point of the essay (thesis statement);
- Each supporting paragraph begins with a topic sentence that addresses one aspect of the thesis statement that requires elaboration;
- The essay's conclusion summarizes the main points discussed in the body of the essay;

	5	4	3	2	1	n/a
Coherence (A clear method of organization: time order, emphatic order, addition, cause/effect, etc.)						

- Supporting examples, details, and evidence are logically organized w/ effective transitions and signal phrases (i.e. first, secondly, then, next, finally, furthermore, in addition, to begin with, as a result, etc.);
- Paragraphs and sentences "stick together" to support or explain the essay's main idea;
- Scientific processes or phenomena are explained/described logically and in detail;
- Explicit connections between ideas, models, etc. are made in the explanation process;

	5	4	3	2	1	n/a
Support (Details: examples, reasons, data, or evidence that explain the main points of the essay)						

- The essay has details and specifics in full support of its main idea;
- Scientific evidence, quotations, examples, data, and models are explained logically and clearly so that they support the essay's point(s);
- Scientific evidence and reasoning are used clearly to support or refute particular ideas or arguments, including those in development;

	5	4	3	2	1	n/a
Content & Synthesis (comprehension / analytical & critical thinking / accuracy & depth)						

- The essay provides a creative & compelling way of thinking about its main point(s);
- The essay recognizes that assumptions have causes and consequences;
- The essay accurately uses scientific laws & principles to draw logical conclusions;
- The essay offers the writer's personal insight to a problem or issue;

Table 9.1. Integrated Grading Criteria (continued)

- The essay presents robust and effective explanations, arguments, or counter arguments in a variety of ways;
- The essay distinguishes between opinion and scientific evidence in contexts related to science, and uses evidence rather than opinion to support or challenge scientific arguments;

Use of Outside Sources (Use & document research sources accurately)						

- A clear line of reasoning links the sources to the writer's own words throughout the essay;
- The essay critically evaluates information and evidence from various sources, explaining limitations, misrepresentation or lack of balance;
- The essay uses just the right amount of outside information, but [it] does not overwhelm [the essay];
- Although it draws from a variety of sources, the essay is clearly the work of its writer: at no point does the reader wonder, "Are these the writer's words, or those of an outside source?"
- The essay uses in-text citation and a variety of signal phrases to weave the sources gracefully, and accurately documents all of the sources used in the MLA or APA style;

Sentence Skills (Mechanics: wording, sentence structure, grammar, punctuation, spelling, etc.)						

- Sentences are varied in length and structure, and are grammatically correct;
- The essay uses appropriate and effective vocabulary as well as correct punctuation;
- Scientific forms of language, mathematical conventions, and appropriate terminology are accurately used to communicate scientific ideas, processes and/or phenomena;

Format (Mechanics: accurate use of the MLA style for both typing and documentation)						

- The essay uses appropriate symbols, flow diagrams and effective graphs in presenting explanations and arguments;
- The essay uses the MLA or APA format correctly for documentation and typing (see textbook for detail);
- The essay is typed or word-processed, double spaced, spell-checked, and proofread;
- The essay uses 12-point font and 1-inch margin on all sides, and has page numbers in the required location.

because things like crop diversification and green manures accomplish the same goals. Crop diversification helps prevent insects and larger animals from damaging the crops . . . [because the outbreak of] the same pest . . . in a field of four or five different types of crops . . . will only damage a percentage of the yield." The CBL experience not only allows him to understand concepts in tangible terms but also shows him how they contribute to sustainable agricultural practices in real-life situations.

CBL experiences like this enable students to witness the benefits of sustainability in real situations. One business major in our class, who was not particularly interested in or knowledgeable about science, became very involved in the work at the food bank farm. He was first drawn to the social equity aspect of the donation of half the food to people in need. He continued to volunteer his time at the farm long after the course was over. Another student, Alice, reflected on the experience in her "course reflection": "This is the only class I have ever taken that has had such a tremendous impact on my life in addition to improving my overall writing & science skills." George wrote at the end of the semester in his "course reflection":

> Before this class doing community service or volunteering was definitely not appealing to me. I didn't see how much could be learned by volunteering just a few hours at the Food Bank Farm. [Yet] I learned methods of harvesting, mulching, preparing vegetables, pest prevention, composting and using green manures. Along with the farming knowledge and conservation of our environment I grew a sense of community. I believe this is just as important, if not more, than the aspects of sustainability. . . . At the farm, I saw people making . . . long-term effort towards sustaining our environment. This gave me inspiration to make efforts myself. There are so many little things like using less electricity in my house, driving less, recycling which I can easily do to be a better member of my community. Volunteering at the Food Bank Farm has given me a sense of pride in my community.

The following year, the New England Adolescent Research Institute (NEARI) was added as a CBL assignment to the course. It was project-based work outside the classroom. NEARI needed help with making its older campus buildings more sustainable. Again, this was a real-life situation that the students could connect with. The director of NEARI, Steve Bengis, "hired" our students as his principal consultants to help him make the NEARI campus greener. This really "upped the ante" for the student work.

We asked those students who chose to work with NEARI to do research projects to educate the director and his staff on ways to become a "greener" campus. Students formed three groups and focused their research on three areas: office supplies, campus food, and campus buildings. All three groups were required to find useful, scientific information, and to present posters to the NEARI staff at the end of the semester. This time, the students sought extra help for their writing, knowing their work would be open to public scrutiny. Bengis (2007) participated in the final presentation of the project. Besides expressing his appreciation for the work completed, he shared the following thoughts with the class (by video):

The wonderful, concrete ideas presented here will bring NEARI closer to a greener campus. It's one thing to tell me that I can get some grant; but it's another thing to show me: "Here is the website for grant application." . . . Putting aerated nozzles on the faucet can be done tomorrow once you told me. . . . So, don't be surprised that 25% of your practical suggestions here will be in place a year from now when you come and visit us again.

This project has shown a new model of interface between the college and community-based nonprofit organizations. . . . Greenizing NEARI is something that we've wanted to do for a long time. We've fooled around the edges, but never had enough resources to do the basic research and come up with a blueprint. Your project has used the resources at the college while making it possible for us to envision a greener NEARI campus. I hope you all have benefited from this project just as much.

What you've presented here today will make a difference in the real world, however small that difference might be in the beginning. It's important for you to know that. . . . I wish I had this kind of experience and opportunity when I was in school.

This kind of community partnership helps show class members that without the scientific background they would not have been able to contribute to this project. And equally, without the skills to adequately communicate their findings, their work would not have had the impact it did. "Being in a hands-on environment and experiencing sustainable practices such as agriculture and green building gave me a head start on becoming a more environmentally cautious individual. Without real-life examples, I might not have taken as much information from this class," said Carol.

The integration of CBL projects into the course content provides a sense of purpose to students' learning of science as well as a sense of ownership in their learning process. For instance, the essays from the early part of a semester tend to report on what they read instead of taking a position on what is being said in the readings. Here is an example of those early seminar papers on the readings:

Bill Bryson makes a point to describe all the heinous creatures and afflictions that can befall any unassuming hiker. . . . His light-hearted observations are punctuated with information about the delicate nature of the trail, and the precarious balance in which it now finds itself. According to Bryson, trees, creatures, and even the very trail in its entirety, are in danger of being lost. He observes how pollution, neglect, and even naturally-occurring phenomena like insects have affected the landscape to extreme degrees.

Miller also speaks of natural resources in their finite condition. He tells us how our natural resources are not as a mechanism . . . but an ever-changing source that can be both sustained and over-used. It is up to mankind, he says, to make the distinction between utilizing and wasting our earth's resources.

Both our readings give a very strong message that our mother Earth is a lot like our own mothers—they may give birth to living things, but they cannot do so at such a rapid rate that can supply on constant demand. And it is up to us to find the balance. (Mary)

Even though the closing analogy is thoughtful, Mary mainly reports on what Bill Bryson and G. Tyler Miller have said, with little of her own reflection. The structure of her paper, in fact, physically separates English readings from science readings, which may indicate that she is still thinking of sustainability along the disciplinary lines.

In contrast, seminar papers toward the end of the semester tend to synthesize the readings and delve into the complexity of sustainability. Kathryn Daley's seminar paper (2005) on Judy and Michael Corbett's *Designing Sustainable Communities* and Miller's chapters takes a stance on dogmatic approaches to sustainable development:

> Though at times we must be conservative and stingy with resources, other times we should—in a good way—be liberal and wasteful. For instance, yes, we must conserve water, but we must also press that saved water into useful service, to share it freely and generously, and like any resources to which we have access, use it with munificence to generate ample, plentiful abundance for as many people as possible. This principle, in many ways, sums up the entire premise of sustainability for me, and I believe for the authors also. This, along with redistributing the decision-making powers by either decentralizing or concentrating them, along with maintaining both diversity and singularity, we can see by example that the Valley Homes creators have figured out an equation equal to success in establishing long-term environmental affluence for their community.

The focus in Daley's paper has clearly shifted from relaying information to making sense of the information. Structurally, her paper discusses three aspects of design principles for sustainability—energy use, diversification, and conservation—each in a paragraph, but never separates them into science paragraph or society paragraph. Just as the challenges facing sustainability are multifaceted, the discussions of sustainable alternatives must be multidisciplinary. Toward the end of the semester, students are becoming increasingly comfortable with making this kind of connection across disciplines.

Curriculum Design

The curricular implications of this interdisciplinary LC model are twofold. Intrinsically, a theme-based LC needs to contextualize its core concepts and skills, and extrinsically, theme-based courses such as this course on sustainability can add viable curricular program options to a general education curriculum by leading into area studies programs.

An interdisciplinary LC like Sustainability should identify and contextualize its core issues and skills if it is intended to broaden the student's understanding of them and achieve any interdisciplinary bridging. But because of the dynamic nature of and contemporary interest in the topic of sustainability, identifying the core content and skills can be just as challenging as contextualizing them. In 2002, global warming was still under suspicion in the political arena. But over the five-year period when the course was offered, the focus of that debate has shifted to what it might take to attain sustainable development. Since the course is committed to learning science and practicing writing in "real-life" situations, course content needs to be adaptive in order to remain relevant to the ongoing discussion of sustainability.

Second, the integrated approach is a work in progress at best to the faculty and students. The goal of bridging disciplinary divides has been under intense examination at the beginning of every semester. Students wondered why the theme cannot be taught in two classes—one on science and the other on reading and writing—about sustainability. What are the rewards of learning the subject in an integrated manner? In hindsight, clearly we designed our course initially with an ingrained disciplinary mind-set despite our conviction that the theme of sustainable development naturally invites the participation of both sciences and humanities.

For example, when we first offered the course in 2002, we assigned 27 of the 28 chapters of *Living in the Environment* by G. Tyler Miller for science reading. They indeed provided both a comprehensive overview and detailed discussion of the topic. Likewise, we went through five of the nine rhetorical modes in John Langan's *College Writing Skills* because we wanted the students to experience as many modes as possible. Along with those chapters were another 170 pages of exercises and drills on thesis statement, topic sentences, use of evidence, grammar, and MLA requirements. Again, we believed that students must know all of this in order to write effectively.

However, it became painfully apparent before long that we were teaching the course in the same way as we would have if we had taught it alone. Students clearly approached the course as two separate entities. We often heard them say in class: "now we're doing the science assignment" or "this is the English homework." Our vision of an integrated approach to sustainability was disappearing even though we still had integrated writing assignments such as "[a]pply the concept of 'positive feedback loop' or 'negative feedback loop' to a real-life situation that you have read about outside the class" (Miller, 2002, p. 51). But integration was more or less in the course structure only. Students were basically left to their own devices for developing interdisciplinary and integrative skills. Better-prepared students were able to pick out what to focus on and how to take advantage of writing as a tool of reflection, while less prepared students were constantly struggling to make connections between science and writing assignments.

The constraints of covering basic concepts and skills, which propelled us to try an integrated approach in the first place, now drove us to reexamine our course design and pedagogy. To help first-year students make meaningful connections to science through practicing basic writing skills, the course would have to move away from broad coverage to in-depth exploration in its science material, and from comprehensive survey to targeted practice in its writing assignments.

By 2004, the retooled course zeroed in on only three aspects of sustainability: sustainable agriculture, green building practices, and energy in its thematic exploration, and four "universal elements" in its writing practice. We chose the three aspects not only for their importance in the discussion of sustainability but also for their familiarity to the students and their interconnectedness. We chose the four "universal elements"—thesis statement, topic sentences, use of evidence, and MLA documentation style—primarily for their usefulness across rhetorical modes. Once the core content came into focus, the course became more intentional in the application of "contextualizing" and "problem centering" as its integrative strategies.

"Contextualizing," as Svetlana Nikitina and Veronica Boix Mansilla (2003) define it, is "an *external* integrative strategy that places scientific and mathematical knowledge in the context of cultural history and the history of ideas" (pp. 11–12). Our course uses it to situate the current environmental calamity in the context of culture and the postindustrial mind-set. From viewing agriculture as the cradle of human civilization to regarding it as the first stage of environmental degradation (Quinn, 1992), or from seeing the environment as a source of unlimited resources to exploit to accepting it as undefined space to enjoy (Hoff, 1992), the "facts" remain the same while the mindset changes. Putting today's postindustrial views of nature in historical and cultural perspectives offers students alternatives in their thinking about the environment. For example, since *Homo sapiens* began altering the environment with agriculture, the production, transportation, and consumption of food have literally determined the landscape of the Earth. Similarly, traditional buildings with thicker walls and larger windows facing south use much less energy than today's utility-dependent houses while maintaining the same level of comfort. The laws of thermodynamics show that the so-called efficiency from the Industrial Revolution is no more than a dramatic increase of energy input in today's products and services. Illustrating the current environmental calamity in those historical and cultural terms can lead students to question many of today's ways of living. But what may be some of the viable solutions? Searching for testable hypotheses and feasible solutions, our course employs the other external integrative strategy of Nikitina and Mansilla's, namely, "problem-centering" (p. 7).

"Problem centering" opens the course to disciplinary tools, in this case, from English, in testing hypotheses and discussing possible solutions through writing. Similar to contextualizing, this integrative strategy can present one problem "with multiple points of entry" as Nikitina and Mansilla have observed in their research, "and thus foster broad *external* connections between the sciences and other fields" (p. 14). But in a different approach from that in their observation, the problem-centered learning in our course intentionally "borrowed" disciplinary procedures and skills from college composition to better comprehend scientific laws and concepts instead of employing "the scientific/mathematical tools and procedures" for "a plan to resolve a complicated environmental issue" (p. 15). Of course, such transferrable exchange of disciplinary tools and skills is not mutually exclusive. On the contrary, it encourages students to approach academic challenges holistically by thinking outside the epistemological confines dictated by academic disciplines.

To practice how to think logically and to deepen the understanding of one's own learning process, the course's writing practice hones in on the four "universal elements" of composition, thesis statement, topic sentences, supporting evidence, and

MLA documentation style, instead of covering a given number of rhetorical modes. We encourage students to put down their thoughts on paper (a rough draft) as much as possible. If they like what they have written down, we instruct them to add a statement at the beginning, which not only summarizes what is on paper but also asserts a position (thesis statement or statement of a hypothesis). Next, the opening statement must be supported point by point with a topic sentence for each ensuing paragraph (use of topic sentence). And each paragraph must have relevant examples, statistics, stories, research data, and so on to logically substantiate a paper's claim (use of evidence). Last but not least, students proofread their essays and retype them in MLA format. This is the writing process that *College Writing Skills* advocates in chapter 2 (pp. 23–38). After two or three assignments like this, most of the students exhibited some degree of confidence using writing to sort out their ideas. Whatever a problem happens to be for a course unit, more and more students analyzed the problem, organized their thoughts, and articulated their arguments on paper. The course becomes increasingly integrated as the comprehension of the subject is enhanced through writing used for an authentic purpose rather than an intellectual exercise.

Faculty Development

Team teaching in a fully integrated course has also constituted a valuable opportunity for faculty development, as both instructors are able to reexamine the importance of their course materials and reassess the quality of their own interaction with students. Unlike peer evaluation or review, team teaching is an experience of self-reflection in the spirit of collaboration. Here the self-assessment is motivated by the desire to contribute more to the course, not to prove anything to anyone.

Kate's Reflection

As a science teacher, I have always stressed the importance of good writing. Whenever students were asked to write a research paper or a lab report, I would correct the English portion of their papers. However, I weighted the science content of their work much more heavily, maybe because of my comfort level with the content assessment or maybe because of the lack of tools on my part for articulating the distinction between a piece of effective writing and an ineffective one I always felt like I was "winging it" for the writing. Teaching with Xian introduced me to the framework of a grading rubric. This dissects a piece of writing into the various components: structural unity, synthesis of material, content, grammar, MLA style, and so on. While it has its limitations, it offers meaningful stepping-stones for both students and me to analyze writing in measurable terms. Now I am able to evaluate more than just the content in a piece of writing.

The grading rubrics have also promoted collaboration between the two instructors. We both graded all written assignments in the course. We asked students to hand in two copies of each formal essay so we could read and grade them first independently and then jointly. We were often very close on the overall grade of a paper, but divergent in given categories on the rubrics. I tended to be more engaged with the content, while Xian looked more closely at the overall structure. Because of this difference in our disciplinary expertise, lively discussions usually ensued about which component affected that given essay the most. We often taught each other about various aspects of the topic while grading.

Working together has also enriched our choice of reading material for the class. We each have our disciplinary texts. For example, mine was Miller's *Living in the Environment* and Xian selected Langan's *College Writing Skills*. But we wanted the students to have a much fuller, broader view than what these two authors could provide, so Xian and I suggested readings that were specific to sustaining the earth. Xian brought Bill Bryson's *A Walk in the Woods* to the class early in the course because it is a humorous look at one person's view of environmental degradation as witnessed on the Appalachian Trail. *The New Agrarianism* (Frefogle, 2001), one of my choices, was a collection of very well written essays on environmental sustainability that Xian could use as examples of good writing as well as for teaching about issues of sustainability. Each year we reused some of the selected readings and changed others as the field changed and expanded. Reading lists from other disciplines have considerably broadened my own appreciation of contributions to sustainability from nonscience writers and thinkers.

Xian's Reflection

The integrated LC has provided content to ENG101 and highlighted the importance of practicing the essentials in composition. Indeed, a piece of effective writing needs to have a coherent overall structure, valid content, interesting and concise language, acceptable grammar, correct spelling, and MLA compliance. But not all writing elements are created equal. While all of them are important, the "universal" elements need to be identified and practiced, particularly in the beginning of an integrated course.

Years within a disciplinary approach to course work have naturally trained students to think of an integrated course in disciplinary terms. In our case, what is a science reading and what is an English assignment seem to plague how students approached their integrated course work. They tend to think of the thesis statement, topic sentences, use of quotes, and MLA requirements when an essay assignment is given. But when a response paper to a Miller chapter is handed out, they tend to write it without much thought to organization. Kate has pointed out that difference because she used to approach writing assignments with a similar disciplinary frame of mind. "Well, you just want us to tell what we think of that chapter" is a rather typical reply from students when I notice the lack of organization in their response papers to science readings.

Having realized that when students thought of science writing they disassociated the coherent structure from the clarity of written thoughts, I redesigned all the essay assignments so that they required use of thesis statement and topic sentences in all writings for the course. Their response papers then became not book reports about what was being said but rather expressions of what they perceived as a chapter's main argument, because the thesis statement asked them to express their clearly defined opinion right at the beginning.

Another benefit of team teaching is that I can teach as a student myself. It not only creates teachable moments but also shows the class that learning can be enjoyed as a lifelong commitment. For instance, it has seldom occurred to me that a perfectly structured essay could be full of pseudoscience. Everyone is entitled to his or her opinion, right? One of the students wrote an essay on the impact of carbon on the environment. It began with a forceful thesis statement, followed by three supporting paragraphs and a conclusion. Each supporting paragraph began with its own topic sentence, and each paragraph used examples and quotes indicting carbon as the devil that threatens to end all lives on Earth. In terms of the overall organization, the essay was well done. To me, it was an A paper, if the content was acceptable. But to Kate, its all-negative views on atmospheric carbon dioxide were questionable at best. Even though CO_2 is present in the environment, neither the student nor I have thought much about it. In fact, all the media hype about CO_2 as a pollutant led me to the misassumption that no attack on the detrimental effects of CO_2 could be overdone. Thus, the paper's position supporting the elimination of carbon dioxide from the atmosphere did not sound an alarm in my evaluation of the paper. Kate, however, pointed out the flaws in the student's scientific thinking. The 0.036% of CO_2 in the troposphere propels the global gaseous cycle and serves as "a key component of nature's thermostat" (Miller, 2002, p. 91). Therefore, eliminating carbon dioxide is ill conceived, if at all possible. In fact, "even slight changes in the carbon cycle can affect climate and ultimately the types of life that can exist on various parts of the planet" (Miller, 2002, p. 91). In other words, CO_2 should be thought of in dialectal rather than dichotomous terms.

My new understanding of the vital role of CO_2 in the environment did not just change the paper's grade but became a teachable moment for the class. We asked the class to reread the chapter on the carbon cycle in Miller and list all the functions that carbon dioxide serves in the environment. For the negatives, the class listed it only as one of the primary greenhouse gases that contribute to global warming; and for the positives, the list went on: all organisms are carbon based, it serves as nature's thermostat, and photosynthesis largely depends on it! The class sat there, just as when I first realized my misassumption about carbon, speechless. In this symphony of ecology, every plant, organism, or gas has its own role vital to the health of the planet Earth. Nothing is "all bad" or "all good" in this biotic context. But why is capping atmospheric CO_2 a top priority in the current efforts for sustainable development? This time, the responses were no longer "good" gases versus "bad" gases; rather development modeled in the industrial-utilitarian mind-set disrupted the gaseous balance in the atmosphere. That disruption took at least two major forms: by reducing the Earth's ability to absorb CO_2 through deforestation, and by putting the "buried" CO_2 back into circulation with fossil fuel consumption. Carbon dioxide should not be blamed for global warming, but human activities that have largely contributed to the imbalance of various gases in the atmosphere, which would lead to drastic climate change on Earth.

An A paper is not only well structured but also must be scientifically sound. Everyone is still entitled to his or her opinion. However, some opinions are more informed by the facts than others. In the field of science, both structure and content must be present in a piece of effective writing.

The impact that the Sustainability LC has had on students, as partly demonstrated by the postcourse activities of Jane, Carol, and Debbie, and the interest it has generated across campus suggests that theme-based courses such as sustainability, diversity, and gender equality can be offered as one of the liberal arts options when students prepare themselves for a life in the 21st century. David Orr (2004) has noted the fragmentation of the modern curriculum as one of the myths about higher education today: Our "hermetically sealed" disciplines and subdisciplines graduate "most students . . . without any broad, integrated sense of the unity of the things" and "routinely produce economists who lack the most rudimentary understanding of ecology or thermodynamics" (p. 11). As a result, "we have fooled ourselves into thinking that we are richer than we are" while none of "the costs of biotic impoverishment, soil erosion, poisons in our air and water, and resource depletion" appear in our gross national product (p. 11). While the accuracy of such an indictment is debatable, the facts remain that advanced degrees did not prevent the Holocaust from happening, and they have not helped even slow the continued destruction of the environment. Yet most of those in charge—government officials, educators, religious leaders, scientists—are the products of our disciplinary education that is in fact "hermetically sealed from life itself" (p. 11). From this perspective, the integrative approach in education may help bring some value and wisdom back into isolated sciences and concepts. In 2008, HCC established the liberal arts and sciences curriculum's Sustainability Studies Program to strengthen its general education curriculum and further enhance all disciplinary studies related to the environment.

Both quantitative and qualitative classroom data collected over seven years underscore the effectiveness of this integrative approach to science education in student learning, curriculum design, and faculty development. For example, we see students apply scientific knowledge in a community service learning setting, thus "field testing" their understanding of fundamental scientific principles. The course curriculum is dynamic, changing in response to the needs of students as they respond to the demands they encounter in the community. As students partner with the community, LC faculty partner with each other to create a "third course"—the interdisciplinary connection to the real world. Understanding the opportunities and challenges of this interdisciplinary LC model may help science educators develop a more integrative approach to introductory science courses in the general education curriculum.

References

Bengis, S. (2007). Address. Final Project Presentation of LC101.01. [Video]. Holyoke Community College, Holyoke, MA, Dec. 10.

Bryson, B. (1998). *A Walk in the Woods—Rediscovering America on the Appalachian Trail*. New York: Broadway Books.

Corbett, J. and Corbett, M. (2000). *Designing Sustainable Communities*. Washington, DC: Island Press.

Daly, K. (2005). "Decentralize, Diversify, Conserve." LC102 seminar paper 5, Holyoke Community College. Holyoke, MA, Oct. 12.

Frefogle, E.T. (ed.). (2001). *The New Agrarianism: Land, Culture, and the Community of Life*. Washington, DC: Island Press.

Hoff, B. (1992). *The Te of Piglet*. New York: Penguin Books.

Langan, J. (2000). *College Writing Skills*. 5th ed. Boston: McGraw Hill.

Miller, G.T. (2002). *Living in the Environment*. 12th ed. Belmont, CA: Wadsworth/Thomson Learning.

Nikitina, S. and Mansilla, V.B. (2003). *Three Strategies for Interdisciplinary Math and Science Teaching: A Case of the Illinois Mathematics and Science Academy*. GoodWork Project Report Series 21. Cambridge, MA: Harvard Graduate School of Education.

Orr, D. (2004). *Earth in Mind: On Education, Environment, and the Human Prospect*. Washington, DC: Island Press.

Quinn, D. (1992). *Ishmael: An Adventure of the Mind and Spirit*. New York: Bantam/Turner.

10 Pedagogies of Integration

Richard A. Gale

ABRUPT CLIMATE CHANGE: What is it and how do we know it when we see it? Why do we call it abrupt when it takes so long to happen? What makes something abrupt rather than gradual on a planetary scale? Where did we get these data, and how do humans figure into this? These were only a few of the questions raised by Carleton College freshman students one September morning. As class discussion shifted from the younger dryas event (which was part of the course) to Greenland ice cores (which was sort of part of the course), to the rise of stable agricultural societies (which was not part of the course), to the domestication of livestock (certainly not part of the course), students began comparing time lines learned in high school to anthropology textbooks nestled in backpacks. Not every question was answered, and not every answer was definitive, but for just under an hour these twelve students were using the knowledge gained in class, matching it with information from other classes, relating it to vacations they had taken and newspaper articles they had read, and committing themselves to outside unassigned research in an effort to make sense of their world and satisfy their own growing curiosity. This was a day of integrative learning, and it was the result of an intentional pedagogy designed to help students weave the often disparate elements of their college years into a fabric functional enough to wear yet fine enough to show.

It was early in the fall of 2005 and the course had only met a few times, but already the students were starting to get into the idea of abrupt climate change as something more complicated and complex than headlines and high temperatures. Because this first-year interdisciplinary science course fulfilled a general education requirement the students formed a fairly representative sampling of the community at Carleton College, but the course was far from typical. In fact, it was part of a scholarly study designed from scratch by two chemists, team-taught by one of those chemists and a psychologist, and never before offered at any college anywhere. The Carleton professor was Tricia Ferrett, one of 21 Carnegie Scholars working to study and understand integrative learning during the 2005–2006 academic year. Her plan was to document and analyze "integrative moments" as they arose in the first-year seminar, and to try to come to some understanding of how students connect prior ideas and learning with the seminar on complex systems. Specifically, she was interested in "the ways students [are] 'going beyond' as they make integrative moves in an inquiry seminar that circles a single transdisciplinary concept—abrupt change—with richly related perspectives from science and social science" (Ferrett, 2006). To get at this question she chose readings, planned discussions, built assignments, and tried to create "an open playground for inquiry" (Ferrett, 2006). In other words, she formed her pedagogy to meet the needs of her students and the objectives of integrative learning. This seems simple enough in the abstract, but it involves a complicated selection process, a dynamic tension between disciplinary content and departmental coverage, curricular planning and emergent awareness, the requirements of the course and the opportunities of the moment.

Integrative Learning

Much has been written about integrative learning and its goal of helping students make connections between isolated course materials, diverse course offerings, classroom knowledge and life skills, the world of school and the world of work. At the outset of their three-year collaboration on the Integrative Learning Project, the Association of American Colleges and Universities (AAC&U) and the Carnegie Foundation for the Advancement of Teaching defined integrative learning as occurring "in many varieties: connecting skills and knowledge from multiple sources and experience; applying theory to practice in various settings; utilizing diverse and even contradictory points of view; and, understanding issues and positions contextually" (2004). This perspective owes much to the groundbreaking work of *Greater Expectations: A New Vision for Learning as a Nation Goes to College*(AAC&U, 2002), which frames "the education all students need" in terms of "the intentional learner" who is both purposeful and self-directed in approaching college, and self-aware about the learning process and the uses of education. "Intentional learners are integrative thinkers who can see connections in seemingly disparate information and draw on a wide range of knowledge to make decisions," the authors note. "They adapt the skills learned in one situation to problems encountered in another" (p. 20).

For most teachers and academics committed to integrative learning, the student is exactly at the center where she should be; it is the student's development, her capacity for meaning making, her cultivation of the skills and abilities to make coherent connections, that matters most. But this learning does not just happen for students. Mary Taylor Huber and Pat Hutchings point out that "integrative learning may also require scaffolding that extends beyond individual courses," and they mention portfolios and self-assessment rubrics as relevant strategies for students trying to make sense of an often-baroque college experience (2004, p. 8). Frequently this is addressed through curricular and cocurricular activities and redesigns and cross-campus initiatives, all well considered and many very successful. However, in the drive to help students develop integrative habits of mind, through freshman seminars and capstone courses, study abroad and service learning, we sometimes forget that a curriculum and a portfolio are only as effective as the pedagogies that support them. For while we are continually asking students to be more conscious of their own learning, we should mention as well that such consciousness is also required of faculty, who must be ever vigilant about offering "better opportunities for [students] to connect their learning within and among courses and contexts" (Huber and Hutchings, 2004, pp. 8–9). Opportunities for integration do not come easily; they require attention to pedagogies that address the goals of such learning, while also accomplishing course objectives. But neither is integration as difficult as it may appear. Through careful pedagogical choices, combining new and established strategies, connecting the unique opportunities of context and content, attending intentionally to ecological changes within and between classrooms, faculty and students can create a more unified and fruitful educational environment where integration becomes more than just an exception, if not the rule. Nowhere is this approach to teaching for integrative learning more important than in the STEM disciplines of science, technology, engineering, and mathematics, where connection is the coin of the realm and understanding is not only improved through but often based upon students' ability to link knowledge and experience creatively, constructively, and coherently. As a recent three-year STEM initiative organized by Project Kaleidoscope suggests, in order to "create the innovative and complex thinkers of the future" campuses must concentrate "on the creation of integrative learners, faculty and students with the skills and confidence to work at the interfaces between disciplines to address both research questions and complex societal problems" (Kezar and Elrod, 2012, np).

But what are the teaching and learning processes that lead students toward (and perhaps even into) integration? According to Julie Klein, "there is no unique or single pedagogy for integrative interdisciplary learning" because all approaches "draw from multiple perspectives on a complex phenomenon for insights that can be integrated into a richer, more comprehensive understanding," and unlike interdisciplinary studies, "in integrative learning, perspectives emanate from disciplines, cultures, subcultures, or life experiences" (2005, pp. 9–10). Klein offers her own "core capacities" for integrative learning, and mentions large-scale approaches to the issue (team teaching, linked courses, core seminars), but provides only a few course-level strategies, including a thematic or problem-based focus (familiar to most STEM disciplines), collaborative projects (explored several times in this book), and case studies (pp. 9–10). And this is perhaps at the heart of the question, for most institutions committed to integrative learning address the issue explicitly from two sides, while ignoring the middle. On the one hand, there are the institutional outcomes of integrative learning, which many campuses seem to embrace. On the other hand is the "array of powerful educational experiences" that makes up the "connective tissue" of integration (DeZure et al., p. 26). In between questions about "what is integrative learning?" and "how is integration fostered on campus?" lies the difficult issue of "how does one teach for it, to it, or with it?" And this is the question that leads to pedagogies of integration.

Integrative Pedagogies

Of course, there is no simple answer to the question of how one teaches to, with, and for integrative skills, but there are strategies that might prove useful to faculty approaching the challenge of fostering integrative learning in all disciplines, especially those interested in connected science. Some of these approaches are strikingly familiar, for there is no rule that says a pedagogy of integration cannot also serve another master or originate from established, conventional knowledge. In fact, many of the pedagogies that support integration also buttress other higher-order skills like critical thinking and the development of moral and civic understanding; indeed, most student-centered strategies "represent models for teaching that if used well can support deep understanding, usable knowledge and skills, and personal connection and meaning" (Colby et al., 2003, pp. 133–134).

In recent years, George Kuh and the AAC&U have been studying and promoting high-impact practices that have been successful at helping postsecondary students learn; this work builds out of the initiative known as Liberal Education and America's Promise (LEAP). The author notes, "Probing data collected through the National Survey of Student Engagement (NSSE), shows that the practices the LEAP report authors initially described—with self-conscious caution—as 'effective' can now be appropriately labeled as 'high-impact' because of the substantial educational benefits they provide to students" (Kuh, 2008, p. 1). These include first-year seminars and experiences, common intellectual experiences, learning communities, writing-intensive courses, collaborative assignments and projects, undergraduate research, diversity and global learning, service learning and community-based learning, internships, and capstone courses and projects (AAC&U, 2008). Many of these practices are, if not commonplace, at least prevalent in the sciences, and some are particularly well suited to integrative learning. Nor is this list exhaustive, for there are other pedagogical strategies (such as problem-based learning, for example) that are clearly integrative and have been demonstrated to have significant impact on student learning. To list all pedagogies of integration would be impossible, in part because in the hands of a deft and committed teacher virtually any pedagogy can be made integrative. But many of the most prevalent approaches seem to fall into four categories:

old friends in new places; new places for new ideas; new ideas for new vehicles; new vehicles as old friends.

Old Friends in New Places

The phrase "old friends in new places" comes from the title of an essay by Joseph Wood Krutch, wherein he refers to the recognition of something familiar in a new environment, in his case, the friends were birds. Here it is linked to the application of tried and true pedagogies to new areas of inquiry, new avenues of thought, new disciplinary and transdisciplinary connections, new course material. At the heart of this category is the idea that a well-constructed course with reasonable assessments and clearly defined outcomes need not be completely redesigned for the purposes of integrative learning. Indeed, many of the strategies we know and love can be made to serve the needs of integration and improve the connective capacities of students; the key is not to revise so much as to reframe. Of course, there is an additional issue at play in this category—the question of workload, which is always a part of any call to innovate pedagogically. Because "faculty perceive the attention their courses get from students to be highly constrained, which makes every moment of classroom and homework time extremely valuable," imposed "curricular initiatives that feel like marginal add-ons are unlikely to rise to a level of importance that justifies attention to them in place of existing content" (Bierman et al., 2005, p. 19). But if large-scale goals promoting integrative learning are achievable through familiar pedagogies, then they are more likely to be implemented.

One example of this comes from Philadelphia University, which began in 1886 as an institute for textile engineering and design and has since grown into a private, professionally oriented institution with majors from architecture to international business to fashion design. When it shifted its focus and changed its name, Philadelphia University also took on the task of merging a liberal education core curriculum with a variety of professional training programs: "the primary challenge . . . [was] the compartmentalizing strategy which our students use to mentally organize their learning experiences and which the curriculum, the faculty, and the campus culture sometimes reinforce" (Frostén et al., 2006, np). In an effort to overcome this compartmentalization and integrate these two course sets and objectives, faculty and administrators sought new ways to make use of familiar and time-tested pedagogies, beginning with a common first-year experience and preexisting senior capstone, both of which operate through the School of General Studies. They introduced a new integrative framework into the first-year course, which already had an annual theme and cocurricular components, in an effort to shift the attention of ongoing pedagogies toward integrative learning outcomes. Additionally, elements of the framework were brought to the capstone and to a project, already underway, for a junior-level transition course. From these modest and generally familiar beginnings, the entire general education model has been reenvisioned with integration as a prominent feature.

In some cases, the implementation of the framework involved rethinking ongoing assignments, in others the expansion of final projects. For example, faculty introduced a focus on liberal education and professional integration into a capstone course on contemporary perspectives exploring current global trends and issues. This emphasis manifested most strikingly in the culminating research project where students were to link a specific global issue to their own career field through a seminar paper with three components: major trend analysis, examination of that trend's impact on the student's field, and a case study of how the trend is affecting the field in a foreign country. This yielded, in one case, an examination of worldwide water scarcity and pollution, combined with increasing pressure on the textile industry to reduce water use and pollution, leading to a case study of how one textile manufacturing process is being changed in Turkey. Another example of the focus, on a smaller scale, comes from a problem-based learning assignment where students must use their professional skills to address a real-world need, such as the design by an architecture student of a cheap, portable, effective HIV-AIDS clinic to address specific needs in South Africa (Frostén et al., 2005). These projects and processes emphasized building from what was already in place, assignments and strategies that already had high impact in their respective environments, while expanding the possibilities for collaboration and critique between general studies and the professional programs. Obviously, this involved stretching faculty into new areas, for what is familiar to a professor of architecture might not be so commonplace for an English teacher in general studies. But the process also included collaborative creation, foregrounding the areas of familiarity, supporting the expansion of practice into new areas, and building slowly enough to ensure ownership and excitement from all those involved in the training of liberally educated professionals.

Similar examples can be found elsewhere in this book, from more traditional science venues. Matthew Fisher at Saint Vincent College describes in his chapter on public health and biochemistry how students are learning the underlying basic concepts of the field through capstone presentation. This traditionally successful high-impact pedagogy, usually course-centric and not always integrated, is morphing into an outward-looking project linking basic disciplinary principles to public problems and practical solutions. It provides students with unique opportunities for integrative connections, not only between course content and the public sphere but also between civic ideas and a variety of scientific perspectives. Additionally, this pedagogy places authority for topic selection and integrative analysis into the hands of the students themselves, offering a level of agency that nonintegrative practices would be hard-pressed to replicate.

The training of legal professionals provides yet another example; at the University of Denver Law School, Roberto Corrada wanted to help his students understand the intricacies of labor law. He used an established strategy common to other disciplines but not necessarily prevalent in his own, the simulation, to provide students with a completely integrated experience of what it was to frame, form, fight for, and ultimately litigate a case through a labor union. In an article for the *Christian Science Monitor*, Mark Clayton (2001) quotes Corrado as saying "one of my goals is to get students to improve their critical-thinking skills. . . . [T]hat isn't necessarily accomplished in a pure lecture format, when they write it down and regurgitate it on a test. For me, they have to create the thing they're learning, work with the subject matter itself to achieve goals—which is what they would do as an attorney." According to Clayton, "Professor Corrada's students were so worked up, they spent hours after class compiling statistics and arguments so compelling that he could not ignore their demand for a more lenient grading curve." But grades were not the only aspect of the course influenced by this familiar pedagogical approach introduced into a new circumstance; through examining this simulation as a substitute for traditional lectures, Corrada learned that students had a higher level of engagement, a deeper level of understanding, and a more active experience of labor law and how it is practiced. Although not framed as integrative learning, it clearly provided such an outcome, and established a precedent for law schools across the country.

Within the world of liberal education there are few pedagogies more highly developed and respected than the academic seminar. Variously defined in terms of size, frequency, content, and practice, it has the generally accepted characteristic of being text-centric and student driven. The seminar is not a common pedagogy in the sciences, at least not at the undergraduate level, and more often than not takes the form of "journal clubs" designed to share and disseminate new work in specific fields. However, in one fascinating case an institution has adopted the seminar and it has morphed into an element of an earth science field course; lectures in the field transform into seminars in the field to allow students to grapple with difficult concepts and develop individual insights as they actively integrate disciplinary content and immediate lived experience. Bettie Higgs's experience with a first-year geology field course is described in Chapter 6 as "navigating wormholes," but in fact it is a perfect example of how a familiar pedagogy can be used in a very unfamiliar venue. The stated goal of the course was the integration of campus-based instruction with field-based experience in an effort to deepen and complicate students' knowledge of geology; simple enough and commonplace for all earth sciences. But this course is different; what was once a frontal data-delivery process of faculty telling students about what they were supposed to be seeing in the field was adapted to suit a more integrative, intentional, and invested model of pedagogical outcomes. Yes, students were expected to learn about geology, and by all accounts they did. But these same students were also invited to engage in the process of discovery, the practice of knowledge creation, and the pleasures of collective understanding. Imagine a group of students working together, using prior knowledge and personal experience to dissect and determine hidden meaning from a core text. What could be more straightforward, more central to the undergraduate experience? But now imagine that the core text is in fact a core sample or strata from a nearby rock formation. The shift in pedagogy might be described as tectonic.

New Places for New Ideas

While having precedents to rely on is always useful, there are times when the desire to promote new learning requires the establishment of new places for new ideas. Thus, there are pedagogies that build opportunities for integration into the learning of new concepts and processes, while likewise placing students into a novel experience or making use of an unfamiliar mechanism. Here experience-based learning strategies are perhaps most familiar, offering students the opportunity to see course work in practice, to manipulate the materials of understanding directly through field studies, to get their hands dirty (physically and metaphorically) with the intersection of knowledge building and the real world. Additionally, there are new technologies and new uses of familiar technologies that provide students with novel venues for integration, and offer teachers the tools necessary to see and connect learning that is happening over space and time. The key to this approach is providing a pedagogical framework outside the traditional course structures where teachers can offer opportunities for integrating experiences and communicating the learning that results.

Perhaps one of the most important connections that can be made in the STEM disciplines is between curriculum and community, specifically the integration of learning through community partnerships. This occurs in many programs, but is not always present within the engineering curriculum. Yet at Ohio University, Gregory Kremer is focusing his engineering capstone design course on what he calls "designing to make a difference" (see Chapter 4). Just as Philadelphia University reconstructed its capstone general education program to incorporate social and public application, this course seeks to help future engineers integrate their learning, knowledge, experience, and aspirations to address the realities of their profession and their professional life. Through community partnerships and capstone experiences, what Kremer calls connected capstones, "intentional connections are made to integrate doing engineering with being an engineer . . . to increase student appreciation of the transcendent purpose of engineering, to provide authentic situations for students to experience the value of professional skills and have structured opportunities to develop and demonstrate those skills, and to maintain or increase the overall quality of the technical work that students complete on the capstone projects." By constructing a pedagogy of integration, combining community context and practical experience, this course connects the science of the classroom with the culture of the profession, thus creating an entirely new landscape for learning and knowing.

Another high-yield venue for this approach is the integration of community-based learning within a learning community. In Chapter 9, Xian Liu, Kate Maiolatesi, and Jack Mino describe the integrative strategies they used to refocus student learning through Holyoke Community College's interdisciplinary learning communities. In a learning community charged with studying sustainability, they employed strategies of contextualizing and problem centering to improve learning in a community-based framework. Learning communities are a tried and true pedagogy as well as a high-impact practice, with Holyoke Community College considered a leader in this approach. Furthermore, community-based learning is quite familiar in undergraduate education, especially in professional programs or at institutions with a service-centric mandate; their use for integration is much less obvious. What's more, the integration taking place in this venue occurs in many forms: connecting classroom and community learning, applying academic tools to real-world problems, linking theoretical knowledge from multiple disciplines, building the skill sets of writing through science and science through writing. This last is of particular interest when considering pedagogies of connected science. As the authors explain, helping students gain ownership of their own science learning was a key feature of the project, and interdisciplinary community-based learning experiences have provided students with a sense of purpose for writing, a reason for learning science, and a connection to local communities. This confluence of science and community within an interdisciplinary context benefits student learning on multiple levels, and placed as it is within the early years of a student's college career, it has the potential for ongoing and lifelong impact.

One final example appears in the writing portfolio initiative at the previously mentioned Carleton College. For many years, Carleton's writing center has been spearheading the use of writing portfolios across the curriculum. For students and faculty using this approach, the portfolio becomes more than a simple repository of work accomplished, more than an assignment in compilation and cursory reflection. Carleton's writing portfolio initiative is an attempt to bridge the work of multiple courses, to link the student experiences of knowledge building and knowledge assimilation to the multiple sites of integration. It is not portfolio as place and package but as crosscutting pedagogy, for the initiative has provided faculty with the skills and tools necessary to build portfolio review into a variety of learning environments and integrating experiences. This effort at integration makes the Carleton project stand out from other portfolio initiatives, for in order to be taken seriously by students, portfolios need to be used coherently and consistently as an integral part of a thoughtful course pedagogy. By putting writing at the forefront of its integrative and liberal learning activities, Carleton has provided a forum for student assessment and articulation of achieved outcomes, while also infusing the campus with a sense of connection with and commitment to the integration of knowledge across all disciplines and experiences (Bierman et al., 2006).

New Ideas for New Vehicles

Often, it is the variety of opportunities and the information itself that sparks and encourages new ideas about pedagogy and learning. In fact, as the educational environment expands into new technologies and innovative forms of representation, there appears to be no end of new ideas for new vehicles. Just as the use of popular film changed the way theater approached the teaching of acting, writing, and scenography, or the advent of fast and simple polymerase chain reaction techniques revolutionized the practice and pedagogy of molecular biology in the lab and in the field, so many of the new media available to faculty are changing on-the-ground opportunities for undergirding student integrative learning. What makes these innovations so exciting is the way in which they empower students to build the kinds of skills we privilege in an academic environment while also addressing the themes and capacities they themselves value. Likewise, new media offer teachers refreshing possibilities for accessing otherwise unavailable student learning, and imagining novel strategies for addressing important outcomes and themes.

One of the most successful and pervasive new media options opening the door for new ideas is the electronic portfolio. Like its "hard-copy" or "paper-based" counterpart, the electronic portfolio is, at least on one level, a repository for student work and a collection or dissemination tool for assessment and evaluation. In this capacity, electronic portfolios offer students the chance to organize, juxtapose, reflect on, and integrate ideas from multiple sources and in multiple forms. Indeed, it is the breadth of opportunity for integration that recommends the electronic portfolio over the more traditional forms, for it can contain more variable material from text and image to video and hyperlinks to the most creative of assemblages built from scratch or gathered from the entire spectrum of a student's experience. In this respect it can serve as a living document, although the word "document" does not do it justice, representing the work attempted and accomplished and shared and even dreamed in a particular course, in multiple courses, and across the academic spectrum of years and institutions. This is in part the vision behind LaGuardia Community College's electronic portfolio initiative, which seeks to provide this vehicle to all students in all courses, creating a pedagogical culture that views the gathering, comparison, integration, and use of information, knowledge, and experience as a ubiquitous feature of instruction.

Since this program began, LaGuardia has expanded the use of the electronic portfolio to the point where now all incoming and outgoing students have, use, and understand the implications of this mechanism. Because the technology is pervasive and supported, offered from the beginning and used throughout, it represents an important link in the chain that is undergraduate education, and stands as a visible commitment from the college and the students. Students now use electronic portfolios to document their academic journey within the context of their life's path, placing course work and career in tandem with personal development and identity formation. Some communicate with friends and family around the world through their electronic portfolios, offering the portfolio's remote riches as a kind of organic postcard or letter from New York that shows and tells the important transitions and transformation a student is experiencing. Likewise, LaGuardia has been profoundly conscious of the faculty development aspect of this initiative, offering faculty not only technical assistance and curricular coherence but pedagogical training in the uses of electronic portfolios. This last aspect cannot be underestimated; as with similar structures supported by a committed curriculum, the key to the electronic portfolio's success is its use in the classroom as a pedagogical apparatus for understanding (Arcario et al., 2006).

Teachers are using this tool in many ways, from the simple to the baroque. It provides an easily accessible and malleable venue for students to gather work in process, show their thinking and doing, "draft" their ideas and impressions as they would, and do through this media, draft a piece of writing or a work of art. Indeed, the electronic portfolio has become a staple of the drafting process for writing and design, presentation and performance, thinking and reasoning critically, integrating the procedures and the products of education. This tool is also quite effective at helping teachers and staff help students to personalize their experiences of course work, giving voice to the latent subtext that often goes unspoken and usually holds the key to true understanding. Likewise, this voicing process leads to reflection, on the work of a course or the work of all courses, as well as the role that academics play in a student's ever-expanding life. One of the staples of undergraduate education, the "compare and contrast" assignment, takes on new life when viewed through the lens, or enacted through the keystrokes and mouse moves, of electronic portfolios. Because of the mediated options, from hyperlinks to digital video to graphic design, the occasions for comparison expand beyond the realm of ideas into the world at large, and the act of contrast can be made more immediate and integrative through visual, auditory, and textual, in its broadest definition, means. Thus, because of the new media, opportunities to connect and combine ideas and learning and meaning making and action expand exponentially in an integrative experience that involves the personal process of creation nested in the public broadcast of impact.

Sometimes the new medium being introduced into the classroom is only new to students and teachers in application and pedagogical innovation. Such is the case with Mike Burke's data rich mathematics classroom (Chapter 5) that utilizes spreadsheets, hardly a 21st-century technical innovation, as a vehicle for integrating mathematics learning into the understanding and interpretation of our everyday world. At the College of San Mateo it is fair to say that many students are familiar with what is considered standard spreadsheet technology, and some may have used spreadsheets extensively in their own work and home lives. But using spreadsheets to better understand different social perspectives with regard to the mathematical concept of function is something of an integrative innovation. Using spreadsheets as a comparative tool to enhance the application of mathematics to real-world problems is a simple, elegant, yet powerful pedagogy of integration. This, coupled with the integration of significant amounts of writing, leads to a confluence of meaning and application that is rare in mathematics classrooms. Burke uses spreadsheets to unpack function, function to affect understanding of social and political and natural phenomena, and writing to complicate and deepen the connections between math and meaning, science and society. It is in many ways the very heart and soul of integrative learning, connected science, and innovative teaching.

Finally there are Gerald Shenk and David Takacs, a historian and environmental scientist respectively who team-teach a course on the social and environmental history of California at California State University–Monterey Bay. Shenk and Takacs were interested in helping their students build a deeper understanding of how California history could be used to understand and engage in political action.

> We share a conviction that many of the injustices in our society are rooted in the history of how Europeans and white Americans have exploited and distributed the resources of the earth. In other words, we believe that social problems are always at some level also environmental problems. Conversely, we hope that our students come to understand that work on environmental issues inherently demands attention to social issues. History, we believe, is not relegated to the past. In our teaching, we see historical understanding as a foundation that helps students become more effective actors in their communities. When they've named an issue of personal political concern—the threats that pesticides pose to community well-being, say, or the dangers that tampons pose to women's health and the environment—students may use history to help them understand how they come to find themselves in this situation, and what they might do to change their communities. (Shenk and Takasc, 2002, p. 139)

This pedagogical perspective was the driving force behind the course, but it also had an influence on the pedagogies adopted and created.

Along the way, Shenk made use of a new pedagogical strategy, the hyperpaper, to help his students develop ever-deepening understanding of the connections and contexts of the history they were learning and the practice they were developing. The idea behind the hyperpaper is similar to that of the progressive paper, only this assignment is best practiced online. Students begin with a single, short writing assignment meant to provide a broad-stroke view of a particular theme or idea. They then, often in collaboration with classroom peers, determine which of the topics briefly dealt with in the paper require elaboration, or are simply of greatest interest to a reader. These topics are "hyperlinked" and provide the theme of the next set of papers. The process continues until students have a multilayered series of documents that, when taken together, encompass the equivalent of a research paper. Along the way they learn the "what" of history as well as the "how" of developing ideas and integrating diverse sources of information. While this new idea could certainly be applied without the use of new media, there are obvious benefits to making this an online activity in terms of tapping into familiar patterns of searching for information, as well as public presentation of ideas and content.

New Vehicles for Old Friends

But the hyperpaper also reminds us that often a new technology, which might seem daunting and opaque to the uninitiated, can sometimes be a repackaging or reconceptualization of an established idea, used to augment rather than replace familiar and friendly strategies; new media can often be seen as new vehicles for old friends. From the simplest online searches, far more familiar to today's students than card catalogues were for those in an earlier time, to the most complex of website designs and creativity programs, which are not unlike the paper- and audio- and video-based strategies of the last few decades, new technologies often build on experiences and activities with logical pre-21st-century corollaries. Furthermore, the ubiquitous connective natures of electronic communications and representations naturally lend themselves to integrative learning, when guided and supported.

Take for example the work being done with weblogs (or blogs) in many courses on campuses around the world. This simple technology, once described as an online diary tool, has in the hands of innovative faculty and students become a powerful tool for connecting ideas and issues to the sources from which they came and the commentaries that illuminate them; it has become a forum for reflection and collaboration, analysis and citation, experimentation and conflict. One example of this can be found in the "asynchronous online discussion forums" being used to investigate students' "Ways of Thinking and Practicing (WTP)" in a core journalism course at Mount Royal University (MacDonald 2010). This course is essentially designed to help students learn how to think like journalists, and is akin to similar courses found throughout higher education dedicated to understanding the tools and perspectives of a particular field of study. Indeed, trying to determine how students learn to think like a physicist, historian, anthropologist, and so on has become something of a closet industry. But it is perhaps most visible in the sciences, because of the connected nature of contemporary inquiry.

Students today face the complexity of science intersecting with politics, economics, local culture, globalization, and ethics and in the process are drawn into the integration of information and understanding within the context of an ever-expanding knowledge base. What makes the blog particularly fruitful in this instance is the way students can directly link their own reading and writing, thinking and speculation, to external and often contentious sources. The process of placement, juxtaposition, and annotation, not to mention the ways in which lines of inquiry develop and change with input from inside and outside the class, makes this an ever-changing and ever-challenging venue for integration. The result is part progressive paper, albeit in electronic form, and part synthesis report, both familiar formats for the presentation of student research.

Just as familiar is the student discussion session, where multiple voices and intellects are brought to bear on a common problem. One reason faculty often prefer discussion to lecture courses is that they provide students the opportunity to make meaning through engagement with ideas that can be challenged and changed within the context of a safe and supportive community structure. Additionally, such courses are often framed as problem based, so as to provide students with an ever-adaptive and largely open-ended research-inquiry experience. But can this kind of collaborative research community operate, cognitively and socially, in a fully online environment? That was the question Kathy Takayama, a biologist at the University of New South Wales, examined in her study that centered on a collaborative yet geographically distant online community that conducted an open-ended genomics research project.

> Research teams composed of five students, each from a different country, collaborated on an open-ended project to analyse, hypothesise, and formulate models based on data obtained from case studies in Human Immunodeficiency Virus (HIV) research. . . . Unlike most courses encountered in their academic careers, instead of being prescribed an aim and corresponding protocol for their experiment, the students' initial (and most difficult) pursuit was "what is my question?" Furthermore, the students were not constrained by the need to derive a correct answer, for the emphasis of VSG was on process via the open-ended learning experience, and not on results. (Takayama, 2004, np)

The pedagogies brought to bear on this new course structure were strikingly familiar, including verbal and textual elaboration of problem-solving techniques and strategies, the use of open-ended—albeit usually asynchronous—communications and conversations, and above all, the application of familiar seminar approaches including community-building activities, collaborative and aggregative discussion protocols, and constant although hopefully not intrusive facilitation by the teacher. All of this seminar-like activity occurred in an uncommon online environment, but with common tools and pedagogies. In fact, Takayama's inquiry bears a striking resemblance to other seminar study projects undertaken in more traditional venues.

Signature Pedagogies

What this suggests, among other things, is that if there are particularly powerful pedagogies of engagement there may also be signature pedagogies of integration. Carnegie Foundation president Lee Shulman has suggested that signature pedagogies share four distinctive features: they are pervasive, routine, habitual, and deeply engaging for students (2005, p. 22). Although the foundation developed the concept of signature pedagogies through its work on professional education, such features also appear in many pedagogies of integration and serve as a useful framework for understanding how such strategies achieve their integrative goals.

Central to the idea of signature pedagogies is their contribution to formation and the development of an identity consistent with our shared notions of behavior, thought, and practice, especially in the professional realm of engineers, doctors, nurses, teachers, and the clergy. One example of this is in the teaching of primary and secondary school teachers, particularly science teachers, who must present and represent science education to future generations. Theirs is a massive responsibility: to craft a message for children that will open the doors of imagination and integration, preparing them to be not only tomorrow's biologists and chemists and geologists and physicians but also the critical and thoughtful citizens who can and will and indeed must understand an ever-changing and more technologically driven society. Many science education programs focus primarily on pedagogical content knowledge and the presentation of effective information; but some reflect an understanding that teachers, like lawyers and doctors and priests, require more than knowledge and data. Teachers hold a special place in our society and require a complex understanding of their unique identity. And this is where signature pedagogies of education become key features of connected science.

One such pedagogy is critical identity development, and it is nowhere more important than in the teaching of future science teachers. David R. Geelan, in his middle school pedagogy, integrates conscious and challenging identity exercises into a science education curriculum (Chapter 7). His questions about what it means to be a teacher, not in the abstract but from the personal perspective of the student, connect science and self, content and concept, pedagogy and perspective in an effort to deepen each student's understanding of what it means to teach science and be science teachers. This is unquestionably a pedagogy of integration, a signature pedagogy of education, and a connected pedagogy of the self.

Another approach that applies to integration also serves as a signature pedagogy of the liberal arts, the academic seminar: "From its nineteenth century roots in German universities . . . to its near-ubiquitous role in the American first-year experience, the seminar has become a much-honored and much-used method of instruction. Originally a forum for advanced graduate students, the seminar is now considered a central feature of undergraduate education and signature pedagogy of liberal learning" (Gale, 2004, np). It is pervasive in that it holds prominence within the liberal arts curriculum, appearing in almost every integrative venue as an important strategy for developing collective and collaborative judgment. Its attention to a particular process of inquiry and discovery through the development of intellectual community and shared understanding is both routine and habitual practice. The seminar requires close reading and broad knowledge, active engagement and connection between not only the ideas and experiences of one student but the thoughts and insights of an entire class. An often disciplinary but just as easily interdisciplinary venue, the seminar integrates information and analysis, text and dialogue, critique and community, while serving as a forum for experimentation and inquiry. In Tricia Ferrert's Carleton College example, mentioned at the beginning of this chapter, the seminar format provided students and faculty with opportunities for engagement and expansion that other pedagogical forums might not have allowed; the class became a vehicle for integrating knowledge, with students making autonomous connections across courses, between experiences, and throughout their own lives. In Bettie Higgs's geology example, it was the seminar that transformed student learning in the field. This pedagogy is based on a communal approach to learning that transcends the assignment, operating at the level of course objectives and educational philosophy; at its best, this is a pedagogy that puts inquiry at the center of the entire course, as a vehicle for the entire class.

Likewise, inquiry is at the heart of problem-based learning. Described as "a pedagogy that asks students to work in small groups to investigate and solve teacher-designed real-world problems in the discipline they are studying," problem-based learning is a favored approach for many disciplines, and can achieve a number of important outcomes from community building to critical engagement, operating at the activity and assignment levels but contributing to the overall structures of the course (Sommers, 2004, np). An important pedagogy of integration, problem-based learning is a way for students to focus attention on common issues, themes, or tasks, working together to share insights and ideas as well as strategies and answers. As students encounter new methods and knowledge within the context of a given course, specified problems provide opportunities to experiment and "play" with existing perceptions in context. In one example from English, Jeffrey Sommers at Miami University–Middletown asked students to engage with the following question: "When reading a book set in the past, how are readers supposed to know what to trust or believe, especially when on some occasions they encounter actual persons, places, events from history and on other occasions are reading about cultures with which they are unfamiliar?" (Sommers, 2004, np). Other approaches to the "real-world problems" of particular disciplines might range from season attendance trends in theater, intermarriage outcomes in biology, or the distribution of pollution credits in economics. Problem-based learning would appear to be most promising as a pedagogy of integration when applied to the gathering of both internal (class-based) and external (real-world-based) knowledge in support or defense of a collective premise and to address a shared challenge.

The importance of real-world integration is nowhere more evident than in emergent pedagogies that grow out of students' need for engagement with an occasion or an issue that trumps whatever might be on the syllabus. This has certainly been the case in recent years, when the *Challenger* disaster, the killing of Matthew Sheppard, the events of September 11, or the Arab Spring rose to prominence in classrooms, demanding attention and consideration regardless of the original course theme or disciplinary focus. But in order to be seen as signature, emergent pedagogy must be not only an incidental approach to a "teachable moment" or "learning opportunity" but also an integral part of a pedagogical commitment to student voice, social engagement, critical inquiry, and integrative learning. Embracing emergence as a course strategy is difficult on many levels, from the logistical to the emotional, but the benefits can be profound in deep student engagement and the sense of ownership at the heart of intentional and perhaps all forms of important learning. Furthermore, the new media environment that has fast become a part of all campus cultures and most students' lives makes it possible to integrate emergent issues into virtually any course, from biology to economics, chemistry to women's studies, physics to music, through active and aggressive use of the Internet and its vast storehouse of useful information and misinformation.

In one instance, a prelaw course on free speech at CSU–Monterey Bay, blogs were used to provide students with opportunities to contribute and comment on artifacts such as articles, images, and commentaries that they believed constituted "evidence" of contemporary questions and conflicts surrounding the issue of free speech. This attention to course content within the context of "breaking news" provided students with a confluence of material and an open-ended occasion for connection, as well as an unexpected venue for their own voice (Reichard, 2004). Another example can be found in a foundation-level science course at Mount Royal University, where Lynn Moorman uses news about natural geological and meteorological phenomena to help students understand how culture and economics influence safety and disaster response around the world. This was brought home for her students quite dramatically when a discussion about the geographical placement of nuclear power plants around the Pacific Rim was followed the next day by a massive earthquake in Japan (Moorman, 2011). Whatever might have been on the syllabus for that week was suddenly of little conse quence, and the immediate concerns of Japanese citizens near and downwind from the Fukashima-Daiichi power plant took center stage for Canadian students in Calgary, Alberta. It is in situations such as these, which all teachers and students face, that the integration of information systems, used with a critical eye and a careful hand, provide immense opportunities for emergent and integrative learning.

In fact, the now ubiquitous online environment provides multiple opportunities for integrative learning through another signature pedagogy familiar to many faculties: the previously mentioned learning community. Variously described and documented, learning communities offer students a level of coherence and overlap which stand-alone courses often cannot achieve. They range from living learning communities that share dorm space as well as class and study time to more limited course pairings that seek to align the pedagogies and outcomes of two or more dissimilar experiences; their strength stems not from the milieu they create but the teaching and learning philosophy and practice they typically embrace. What makes learning communities of any stripe succeed as a persuasive, routine, habitual, engaging pedagogy is their internal mechanism for understanding and their establishment and championing of a seeing, thinking, knowing process that will stand students in good stead, continuing to influence their educational journey beyond the initial term or year of the community. These pedagogical strategies include cross-disciplinary reading and critical review, the assisted analysis of text and artifacts from at least two separate yet triangulated perspectives, and the carrying of disciplinary skills both in tandem and in harmony from one learning landscape to the next. Perhaps most important, learning communities provide unparalleled opportunities for making connection within individual assignments, between course units, and among linked courses. In one study, Jack Mino at Holyoke Community College examined the ways in which students made a variety of conceptual connections by adapting the "think-aloud" protocol for harvesting evidence of student learning into a "link-aloud" that demonstrated evidence of student integration (Mino, 2006).

Finally, the portfolio experience, when approached as pedagogical strategy, may be another signature pedagogy of integrative learning, especially in light of its flexibility and portability. Portfolios have long been used within courses to help students make connections and reflect on continuing and cumulative work. Some programs have embraced them as a way to link the work of often disconnected courses, and entire institutions have turned to portfolios as a way of capping two or four years of study. In one yearlong study of program portfolios in the Hutchins School of Liberal Studies at Sonoma State University, students determined that although useful as a culminating repository of completed work, portfolios were not as useful as they could be because they were not an integral part of the curriculum or of any one course. Their recommendations for improvement included explicit expansion of the portfolio process throughout the program, integration of portfolio learning objectives within all courses, and ongoing revision of the portfolio and its use at regular intervals (Gale, 2001). Like any pedagogy, signature or otherwise, the portfolio must be viewed as not only integrative with regard to student learning but integrated into the learning fabric of the context in which it functions.

With the advent of electronic portfolios and online repositories of student work and experience, opportunities for integration become legion, if these tools for learning are themselves integrated into pedagogical practice. The electronic portfolio is by definition engaging, as it involves a level of commitment, innovation, interpretation, and freedom that encourages active connection. But to make portfolio use a truly pedagogical experience, it must become routine, not just on campus and between courses but within courses, where it must go beyond its role as a place to gather evidence. Routine use and revision of portfolios within courses is the first step, providing the building blocks for habitual use beyond the initial site of connection. In order for this pedagogy to be taken seriously by students it must be used thoughtfully by faculty within the context of short- and long-term learning objectives. For unlike other pedagogies, the portfolio is not automatically pedagogical in its use and application in the classroom, is not always familiar to the faculty, and is not commonly "owned" or managed by the teacher. Faculties working at institutions with portfolios require thoughtful instruction in not only the technologies of operation and access but also the ways in which the philosophy behind the apparatus can be infused into teaching and learning in multiple, connected, and overlapping courses.

All these pedagogies of integration, and many more, share certain qualities and elements that operate regardless of the level at which they are used. They acknowledge the realities of a changing world where disciplinary and curricular isolation are neither feasible nor desirable and in doing so blur the boundaries between areas of expertise, stretching teachers and students into new cognitive and affective arenas. Such pedagogies of integration also embrace a level of expanding ambiguity uncommon and perhaps even uncomfortable in most pedagogical contexts, but nonetheless valuable for understanding the rapid-fire reality of today's students and the ever-changing environment of the modern world. They require intellectual dexterity on the part of the teacher and the student, and the ability to speak to, if not from, a broad spectrum of knowledge and experience. They also embrace a commitment to dialogue between ideas and issues as well as people and positions, as well as an acceptance of conflict that requires the sharing of authority and control within the classroom, especially when the theory of the content meets the reality of the world. As a result of these aspects and many more, most pedagogies of integration necessitate a flexible perspective on assessment, a nimble approach to achieving objectives, and a realistic attitude toward the coverage of content. In other words, pedagogies of integration ask us to develop a culture of inquiry in the classroom that focuses on more situational and often less-than-obvious evidence of student learning. These pedagogies are student-centered, student-influenced, and often student-directed, shifting the center of gravity and balance from coverage and content to engagement and experience.

Beyond Pedagogies

All these pedagogies, and more, provide opportunities for integration of student learning, but all of them also require attention to and instruction in assessment, curricular alignment, and faculty development. It is all well and good to say that we want our students to be integrative learners, but how will this integration be demonstrated and how will those demonstrations translate to the world of grades? Some institutions, such as the University of Charleston, are grappling with this question as it relates to assignments, and to collecting and sharing the kind of faculty work that serves to prompt student integration. Others, like Michigan State University, are looking to special contexts and programs that can demonstrate integration for other campus units. Portland State University, Salve Regina University, and my own institution, Mount Royal University, are including issues of integration and assessment in larger conversations about general education, core courses, and student transfer as they strive for more coherence in their curricula.

The question of curricular coherence and alignment is always important, but nowhere more so than in integrative learning. For if colleges and universities are asking students to make sense of their experience, do they not also then have a responsibility to be sensible about how that experience is framed and supported? This is at the heart of Carleton College's efforts to identify and understand campus-wide crosscutting forms of literacy, but it is also evident in SUNY Oswego's innovative Catalyst Project. Connected science and integrative learning are at the heart of such independent ventures as Project Kaleidoscope, the Science Education Resources Center, and the Carl Wieman Science Education Initiative at the University of British Columbia. And of course, all of this is only possible when faculty are provided with opportunities to connect what they are doing in their own classrooms with the work being accomplished across the hall, the quad, the country, and the globe. Too often teachers are left to their own devices, asked to address overarching objectives and vaulted outcomes without proper guidance or assistance. If integrative learning is only as good as the pedagogy supporting it, then integrative teaching is only as successful as the educational development and institutional vision that makes it possible and then makes it work.

Perhaps this suggests that a focus on the pedagogies of integration is only the beginning, and that we also need an integration of pedagogies in support of deeper, more connected, and more inclusive student learning. In fact, just as integrative learning is the connection of ideas, experience, and inquiry, pedagogies of integration may be best understood and facilitated through collaboration among faculties, administrations, and, of course, students. This is precisely what happened at the College of San Mateo, where a new pedagogical strategy appeared, one that braided long-standing and successful learning communities, attention to the goals and inquiry processes of the scholarship of teaching and learning, and growing administrative interest in and support for integrative learning. This vision started a campus-wide dialogue about integrative learning, attracted new faculty to the idea of pedagogical collaboration, and most important, helped students from different isolated course contexts begin to see and use the links between knowledge and meaning that had previously been unavailable (Ball et al., 2006).

Yet integrative learning, despite its current vogue, is not a new idea. Indeed many assignments ask students to connect material across multiple contexts, frequently course content is provided in coordinated units that seek to draw on real-world examples and student experiences, and linked or otherwise integrated courses have been the norm at many institutions for decades. It is important to acknowledge that while integrative learning may seem both logical and likely given the right institutional context, it is not always easy to develop an intentionally connected pedagogy that will address the needs of students, the hopes of faculty, and the desires of campus administrations. But neither is it an impossible task; much of what already occurs at classroom, program, and institutional levels strives for integration, especially in the STEM disciplines. As Susan Elrod pointed out in an issue of *Liberal Education*, "the complex challenges facing our society in this century—especially challenges related to energy, climate change, and global health—will require interdisciplinary, integrated solutions from a new generation of scientists, engineers, agriculturalists, nurses, teachers, and citizens equipped with the tools to grapple with the social, civic, political, and scientific facets of these problems" (2010, p. 32). The difficulties lie not in imagining integrative pedagogies or in combining innovative strategies for teaching and learning; rather, the key to success is in communication, collaboration, and connection within and between institutions. For one of the most useful lessons learned by campuses that have collaborated on issues of integrative learning is the way that intercampus inquiry and exploration can yield cross-campus pedagogical innovation. When faculty are given the opportunity to collaboratively rethink their pedagogy, when institutions are rewarded for inventive structures that support pedagogical innovation, and when national or even international attention is turned to the questions of and strategies for integrating student learning, the results become greater than the sum of their parts.

While integrative learning is by no means the be-all and end-all of undergraduate education, it is certainly a central feature of liberal learning, a core incubator of academic success and lifelong meaning making, and more than any other pedagogical approach it is at the heart of a truly connected and connecting science education. It has the ability to change the way students see the world and make sense of the often-overwhelming information, knowledge, and experience they encounter on a daily basis; and in so doing, integrative learning gives students the opportunity to change the world they are in the process of creating. Yet in order to make a difference integrative learning must be addressed with intention, become an integral part of a pedagogical commitment. What makes this not only desirable, but likely, is that it is already central to many teachers' ideas of a good education generally and good science education in particular, relying and building on the work underway in many classrooms. In fact, the key to integrative learning and the pedagogies of integration may be found in the realization that faculty not only *should* be doing his kind of work, and *can* do this kind of work, but that they often *are* reinforcing deeper student learning, and *already* teaching in ways that can ultimately lead to connections that matter and persist.

References

AAC&U (Association of American Colleges and Universities). (2002). *Greater Expectations: A New Vision for Learning as a Nation Goes to College*. Washington, DC: AAC&U. Available at http://greaterexpectations.org/. Accessed July 13, 2006.

AAC&U (Association of American Colleges and Universities). (2008). *High-Impact Educational Practices: A Brief Overview*. Washington, DC: AAC&U. Available at http://www.aacu.org/leap/hip.cfm. Accessed June 12, 2011.

AAC&U and the Carnegie Foundation for the Advancement of Teaching. (2004). *A Statement of Integrative Learning*. Washington, DC: AAC&U. Available at http://www.aacu.org/integrative_learning/pdfs/ILP_Statement.pdf. Accessed Nov. 27, 2012.

Arcario, P., Clark, J.E., Eynon, B., and Maodza, N. (2006). *First Year Experience Academies: Reconfiguring the First Year of College through Integrated Courses and ePortfolio*. Stanford, CA: Carnegie Foundation for the Advancement of Teaching. Available at http://www.cfkeep.org/html/snapshot.php?id=6619576883148. Accessed Oct. 11, 2006.

Ball, J., Burke, M.C., and Mach, J. (2006). *College of San Mateo July 2006 ILP Snapshot*. Stanford, CA: Carnegie Foundation for the Advancement of Teaching. Available at http://www.cfkeep.org/html/snapshot.php?id=12680979271122. Accessed Oct. 11, 2006.

Bierman, S., Ciner, E., Lauer-Glebov, J., Rutz, C., and Savina, M. (2006). *Carleton College ILP Snapshot*. Stanford, CA: Carnegie Foundation for the Advancement of Teaching. Available at http://www.cfkeep.org/html/snapshot.php?id=81128863. Accessed Oct. 11, 2006.

Bierman, S., Ciner, E., Lauer-Glebov, J., Rutz, C., and Savina, M. (2005). Integrative Learning: Coherence out of Chaos. *Peer Review*, 7 (4): 18–20.

Clayton, M. (2001). Students in 'Oppressive' Law Class Learn to Negotiate—Fast. *Christian Science Monitor*, Dec. 4, np. Available at http://www.csmonitor.com/2001/1204/p15s1-lehl.html. Accessed July 13, 2006.

Colby, A., Ehrlich, T., Beaumont, E., and Stephens, J. (2003). *Educating Citizens: Preparing America's Undergraduates for Lives of Moral and Civic Responsibility.* San Francisco: Jossey-Bass.

DeZure, D., Babb, M., and Waldmann, S. (2005). Integrative Learning Nationwide: Emerging Themes and Practices. *Peer Review,* 7 (4): 24–28.

Elrod, S. (2010). Project Kaleidoscope 2.0: Leadership for Twenty-First-Century STEM Education. *Liberal Education,* 94 (4). Available at http://www.aacu.org/liberaleducation/le-fa10/LEFA10_Elrod.cfm. Accessed July 13, 2006.

Ferret, T. (2006). *First-Year Students "Go Beyond" with Integrative Inquiry into Abrupt Change.* Carnegie Scholar Final Snapshot. Stanford, CA: Carnegie Foundation for the Advancement of Teaching. Available at http://sakai.cfkeep.org/html/snapshot.php?id=86948187730227. Accessed Nov. 27, 2012.

Frostén, S., Roydhouse, M., and Shrand, T. (2005). Integrating Student Learning and Pedagogies. PowerPoint presentation at the ISSOTL Conference in Vancouver, BC, Oct. 16. Available at http://www.cfkeep.org/html/snapshot.php?id=14500709543471. Accessed Oct.11, 2006.

Frostén, S., Roydhouse, M., and Shrand, T. (2006). *January 2006 ILP Snapshot.* Stanford, CA: Carnegie Foundation for the Advancement of Teaching. Available at http://www.cfkeep.org/html/snapshot.php?id=14500709543471. Accessed Oct. 11, 2006.

Gale, R. (2001). *Portfolio Assessment and Student Empowerment.* Stanford, CA: Carnegie Foundation for the Advancement of Teaching. Available at http://www.cfkeep.org/html/snapshot.php?id=2478563. Accessed July 13, 2006.

Gale, R. (2004). The "Magic" of Learning from Each Other. In *Carnegie Perspectives.* Stanford, CA: Carnegie Foundation for the Advancement of Teaching. Available at http://www.carnegiefoundation.org/perspectives/magic-learning-each-other. Accessed Nov. 27, 2012.

Huber, M.T. and Hutchings, P. (2004). *Integrative Learning: Mapping the Terrain.* Washington, DC: Association of American Colleges and Universities.

Huber, M.T., Hutchings, P., Gale, R.A., Leskes, A., and Miller, R. (2003). *Integrative Learning Project: Opportunities to Connect.* Washington, DC: Association of American Colleges and Universities; Stanford, CA: Carnegie Foundation for the Advancement of Teaching. Available at http://gallery.carnegiefoundation.org/ilp/. Accessed Nov. 27, 2012.

Kezar, A. and Elrod, S. (2012): Facilitating Interdisciplinary Learning: Lessons from Project Kaleidoscope, *Change: The Magazine of Higher Learning,* 44:1, 16–25.

Klein, J.T. (2005). Integrative Learning and Interdisciplinary Studies. *Peer Review,* 7 (4): 8–10.

Kuh, George D. (2008). *High-Impact Educational Practices: What They Are, Who Has Access to Them, and Why They Matter.* Washington, DC: Association of American Colleges and Universities.

MacDonald, R. (2010). *Learning Disciplinary Ways of Thinking and Practicing in Two Introductory Journalism Courses: Evidence from Online Discussions.* Calgary, AB: Institute for Scholarship of Teaching and Learning. Available at http://www.mtroyal.ca/wcm/groups/public/documents/pdf/macdonaldprojectsnapshot03.10.pdf. Accessed November 27, 2012.

Mino, J. (2006). The Link Aloud: Making Interdisciplinary Learning Visible and Audible. Stanford, CA: Carnegie Foundation for the Advancement of Teaching. Available at http://www.cfkeep.org/html/snapshot.php?id=29945016959631 . Accessed September 15, 2006.

Moorman, L. (2011). Interview, Mount Royal University, May 31.

Reichard, D. (2004). *Cultivating Legal Literacy in a Free Speech Class: How Undergraduate Students Develop Deeper Understandings of the Law*. Stanford, CA: Carnegie Foundation for the Advancement of Teaching. Available at http://www.cfkeep.org/html/snapshot. php?id=791. Accessed July 13, 2006.

Shenk, G. and Takasc, D. (2002). Using History to Inform Political Participation in a California History Course. *Radical History Review*, 84 (Fall): 138–148.

Shulman, L. (2005). "Pedagogies of Uncertainty." *Liberal Education*, 91 (2). Available at http://www.aacu.org/liberaleducation/le-sp05/le-sp05feature2.cfm. Accessed on Nov. 27, 2010.

Sommers, J. (2004). *'Based on a True Story': Problem-Based Learning in an Introductory Literature Course*. Stanford, CA: Carnegie Foundation for the Advancement of Teaching. Available at http://www.cfkeep.org/html/snapshot.php?id=76560535. Accessed July 13, 2006.

Takayama, K. (2004). *Final Project Snapshot*. Stanford, CA: Carnegie Foundation for the Advancement of Teaching. Available at http://www.cfkeep.org/html/snapshot. php?id=72200113881566. Accessed July 13, 2006.

PART IV
BROADER CONTEXTS
FOR INTEGRATIVE LEARNING

11 Integrative Moves by Novices

Crossing Institutional, Course, and Student Contexts

Tricia A. Ferrett and Joanne L. Stewart

THE FIRST-YEAR SEMINAR students were not going with the plan. They were engaged in a journey through richer and richer sets of ice core data going further back in time. First 50,000 years before the present, and eventually several million years before the present. They worked to interpret wildly oscillating climate data, drawing out meaning while reading *The Two-Mile Time Machine* (Alley, 2002) and *The Tipping Point* (Gladwell, 2002). Each time they cycled further back in time with more data, they encountered surprises—new data trends, new theories for abrupt climate change, and a need to reconsider their understanding. The instructor was ready to stop the data game and move to the compelling issues of human vulnerability to abrupt climate change. But the students were not ready. Several of them made the case that they still did not have a solid understanding of the mechanisms for abrupt climate change. They noted that at every turn back into deep time, they had rewritten their understanding, sometimes in dramatic ways. "We are not done," they argued. The rest of the class seemed to agree.

Because the instructor was intrigued by this emerging coup, she asked the students what they wanted to do. They requested much more climate data, "going back 4.6 billion years, the lifetime of the Earth." The instructor silently gasped, thinking, "where are we going to get those data? I really don't know anything about global change mechanisms on that time scale. I'm just a chemist. But this is what I wanted, for my students to own the inquiry." She swallowed her fear and talked with the students about which earthly objects they thought would carry that kind of history (rocks, they are really old, right?). Powerful and interdisciplinary questions were leaking out of them. What about mass extinctions and the dinosaurs? The time before plant life? The time when the continents were a huge single mass? The instructor had never witnessed such a groping for the past in students—and such demands by students to structure the course themselves. After class, she called her local global change biologist and started on a new data adventure, one the students would partly define over the next several weeks with small research projects.

Several months later, at a different institution with a different group of students, an instructor observed similar agitation in the classroom. These students, too, had read Alley and Gladwell. They had examined detailed ice core data, studied connections between orbital changes and climate change, and learned about ocean circulation patterns. They were now being pushed to think about the impact of climate change on humans. "You really want us to say what we *believe*?" the students asked again, struggling to understand how their personal beliefs and values had any relevance in this science class on climate change. The question before them involved making a choice: What should be done to reduce human vulnerability to climate change? How could they answer this question, they wondered, when they did not have a "complete picture" of how climate was changing? And how could climate change possibly matter in parts of the world that were struggling with war and poverty? How can the actions of one person make a difference? The students were deeply engaged in the complexity of a very real problem. They were struggling to integrate their emerging understanding of global climate change with their developing identity as citizens in a global community. "But wasn't that the whole point of *The Tipping Point*?" wondered one student, "that little things can make a big difference?"

This was a seminar on abrupt change in climate and human social networks. It was taught by Tricia Ferrett at Carleton College for the first time in the fall of 2005, and then passed to Joanne Stewart at Hope College, to improve, adapt, and teach many more times in the next several years. The Hope context was quite different—the seminar was a general education offering and enrolled students from all four years of college. This seminar deliberately connected students' science learning to their own beliefs and values—an approach aligned with Hope's religious context. In contrast, at Carleton the course centered on having the students think about fast changes in climate, and in how they "changed their mind" over the course.

The main goal for the course was for students—students who were novice scientists—to develop their integrative thinking skills and apply them to a science-rich problem. This raised two questions: What exactly are integrative thinking skills and what might we accept as evidence of them? It also raised a challenge: Can novice science students do this? Is integrative thinking an "expert" skill or can we think about integrative learning as a continuum from novice to expert levels? Could we develop learning opportunities that would allow students to first "dip their toes" in integrative thinking, and then provide experiences to help them develop more advanced integrative skills?

The course that we developed allowed us to explore these questions and challenges. We approached the course as an opportunity for inquiry, asking the research question "What is integrative learning in the context of this course and these students?" The chapter is structured in the way the experiment unfolded. It begins with a description of the development of the course and the literature we drew on to create opportunities for integrative learning. It moves next to descriptions and examples of the evidence for student learning, which consisted primarily of student writing, but included some student data analysis exercises and faculty observations from in-class discussions. The chapter concludes with an innovative *distillation* of the types of integrative thinking or "integrative movelets" we observed in student writing.

The distillate (the authors are chemists, after all) or integrative thinking scheme draws mostly from the analogies that students were making—analogies with varying degrees of sophistication and productivity. The scheme's power and utility comes from the fact that it was abstracted from multiple contexts (different writing assignments) and different students at different institutions. Though we do not claim that all integrative work launches from analogic thinking, we do believe that awareness of the ideas in our scheme can help students be more intentional about integrative thinking—and can help faculty implement processes that can lead novices to higher cognitive forms of integrative thinking.

Course Design for Connected Science Teaching

We designed this interdisciplinary science-rich course with students' futures in mind. They would need to understand more about how to address complex systems and problems. Traditional studies of organisms, rocks, frictionless planes, and simple chemical reactions would be needed by this generation of college students, but it would not be enough. Our students would be confronted with 21st-century issues like climate change, environmental degradation, the consequences of a global economy, and so much more. We thought it was time to shift the science learning environment to reflect all of this. Integrative and creative thinking would need to take a much higher priority, but these skills seemed somehow out of reach for our novice students. What could we realistically expect?

Three frameworks helped shape our thinking. One came from the literature on student intellectual development from William Perry (1999), Marcia Baxter Magolda (2000, 2001), King and Kitchner (1994), and Mary Belenky and colleagues (1986). We also used ideas about "adaptive expertise" (Bransford et al., 2000), which describes innovative, flexible, and responsive thinking. Finally, we drew on the work on integrative learning of the American Association of Colleges and Universities (AAC&U) and the Carnegie Foundation for the Advancement of Teaching (Huber and Hutchings, 2003; Huber et al., 2007).

The literature on student intellectual development provides a foundation for thinking about how to work with novice learners effectively, and describes the types of support they will need to grow and excel. Baxter Magolda, in particular, shows that intellectual, social, and personal development are intertwined, and that teachers must attend to all three to foster student learning and growth (2000, 2001). Therefore, our choice of pedagogies included social learning, through small group or whole-class interactions, and personal learning, through supporting students to make connections between course content and personal experiences or beliefs. Perry describes the ways in which novice learners tend to be concrete, dualistic thinkers (1999), so we started the course with concrete, data-analysis activities, but quickly mixed in readings from Gladwell's *Tipping Point* (2002) in order to introduce the idea of nonlinear, probabilistic thinking. Ideas from "adaptive expertise" (Bransford et al., 2000) helped us articulate the kind of flexible thinking skills we wanted to promote. An adaptive expert views a problem as a "point of departure and exploration." We return to this notion when we describe our problem choice for the course, namely, understanding abrupt climate change.

The literature on integrative learning also helped us describe the type of student learning we sought. A statement on integrative learning provides the following definition (Huber and Hutchings, 2003): "Integrative learning comes in many varieties: connecting skills and knowledge from multiple sources and experiences; applying theory to practice in various settings; utilizing diverse and even contradictory points of view; and, understanding issues and positions contextually" (p. 1). This idea of connecting skills and knowledge across multiple domains or boundaries served as the basis for many of the student writing assignments. For example, we asked students to use Gladwell's framework for social tipping points to describe abrupt climate change tipping points, thus making connections between the social and physical sciences. At Hope College, students were asked to use what they had learned about values in science to rationalize some of the history of the science of abrupt climate change. These challenging assignments required carefully structured activities that first built students' understanding of Gladwell's framework, the science of abrupt climate change, values in science, and the history of abrupt climate change science. We were careful to provide this foundation in order to support the challenging integrative tasks we presented to students.

To be honest, we were nervous about the feasibility of asking our novice science students to tackle these high-order intellectual challenges. We were befuddled by the magnitude of the challenge—not so much by the teaching challenge, which we relished as another research-like adventure with our students. Our own experience and inclinations in science education reform and undergraduate research, mixed with strong leanings toward social justice causes, helped fuel us. What kept us awake at night was the lack of a way of thinking about how our students, novices to both the scientific disciplines and integrative learning, would develop in front of us. What kind of work could they really produce? What would be most difficult for them? What would their learning look like? Could they really learn a lot of good science with this approach? How could we help them?

The final design concerns were our choice of course content and the sequencing of activities and assignments. If we wanted students to learn to view problems as a "point of departure and exploration," then it would be counterproductive to focus on science knowledge that was "done" or had been worked out in the past. Instead, we wanted our students to study "living science"—science that was being built today, science that carried with it real uncertainty and controversy. The science of abrupt climate change fulfilled this role. This choice allowed students to explore complex systems and the role of inherent uncertainty in these systems. Our students would study this content in the context of messy problems and act like real scientists and citizens. We tried to create a learning environment that was rigorous scientifically, yet social, fluid, creative, and integrative—a setting that deepened their science learning on both the content and process sides. To do this, we used a recursive process of heavy data interpretation mixed with scientific theories to explain those data. Almost daily, our students crossed traditional boundaries, delving into the science of social networks, multiple scientific disciplines (we did not name them, we just did it), and their own background and experience.

The Carleton College course was taught as a small first-year seminar designed to introduce students to the liberal arts and college learning. The course emphasized student participation, writing, learning through discussion, intellectual independence, and critical thinking skills. The Hope College course was part of the General Education Mathematics and Science Program. These interdisciplinary courses are designed for nonscience majors, and they seek "to enhance students' understanding of the power and limitations of mathematical and scientific investigation as it applies to real-world questions and problems" (Hope College, 2012, np). At both colleges, nearly all our students had not taken a college science course—so they were novices in that sense. There are important institutional differences, though, that will come into play. Hope College has a primarily regional student population. The students are mostly white Protestant or Catholic and traditional college age, and they come to Hope seeking a Christian college experience. Carleton students come from every state in the United States and many different countries. The students typically have broad academic interests and are often described as being very intellectually curious. Most of them are also comfortable with challenging authority in and out of the classroom.

The courses shared a number of essential features, including much scientific content and the approach to learning science. Students studied historical climate data records and theories to examine the emerging paradigm of abrupt change. This approach allowed for exploration of complex natural systems and the human processes for building science knowledge. As noted above, the inquiry circled through climate data interpretation, going increasingly further back in time. We also shared the reading of Gladwell's *The Tipping Point* so that students could juxtapose their learning about tipping points in both social and climate systems. Our students also read and discussed parts of Alley's *Two Mile Time Machine* (2002) and engaged in similar class activities to examine data and theories for abrupt climate change. In some cases, we shared assignments designed to help students integrate across the boundaries of science and nonscience, science content and science process, and science and personal values and beliefs.

The courses differed in both subtle and important ways. At Carleton, students focused on questions that climate researchers are asking themselves today: How fast does global climate change? Why does climate change quickly? How have humans been affected by abrupt climate change? Is abrupt climate change in our future? A unique feature of the Carleton course was that students also engaged in learning activities with a second seminar, taught by Larry Wichlinski of the Department of Psychology, under the common theme of paradigm shifts in science. Larry's seminar focused on the mind-brain relationship in neuroscience. Jointly, students read not only Gladwell but also Thomas Kuhn's *The Structure of Scientific Revolutions* (1996) and portions of Howard Gardner's *Changing Minds* (2004) and Jared Diamond's *Collapse* (2005). Students examined abrupt changes in the human mind at the level of individuals, social groups, scientific communities, and civilizations. They reflected throughout the course on their own mind changes.

At Hope, a major course goal was to enable students to integrate science into reflective thinking and decision making. This led to a course that focused on a slightly different set of questions: How does science work and what is the "science" way of knowing? Why should we care about understanding science? What is the role of science in decision making? The course approached the first question from several perspectives. Students began by analyzing real data on abrupt climate change. Then they examined the history of abrupt climate change, which illustrated for them the nonlinear pathway by which science often progresses. Finally, they considered the role of beliefs and values in science. In what ways did beliefs and values influence the development of the scientific understanding of abrupt climate change?

Both courses used a historical case study approach to look at the impact of climate change on humans. Through the case studies, it quickly became apparent that the impact of climate change is intertwined with other factors such as environmental problems and sociopolitical issues. This introduced more complexity into an already complex problem, and Diamond's *Collapse* (2005) helped provide a framework for thinking about that complexity. We were not so interested in having the students find "an answer" to how to deal with a complex problem like climate change. Rather, we wanted them to explore the relevant and varied perspectives involved with such issues, in scientific, thoughtful, and reflective ways. We wanted a "pedagogy of uncertainty," aligned with work by Lee Shulman (2005), consistent with our inherently uncertain course content and goals for working with complex problems and systems.

At Hope, the course pushed students much harder to think about the role of science and their own values in real-world decision making. Hope students did a final project that asked the question, "What should be done to reduce human vulnerability to climate change?" They picked a place in the world and used the science they had learned to make predictions about the future impact of climate change on the region. They included in their analysis environmental problems and information about social, economic, and political issues in the area. Their suggestions for reducing human vulnerability had to be clearly supported by their evidence, and they were required to explain how the suggestions were congruent with their personal beliefs and values. In contrast, the students at Carleton continued to circle through the question "How has your thinking about X changed?" The Carleton focus on mind change was a different way to move the students toward a more reflective stance about their learning. Both courses, in different ways, aimed to help students become more aware of their own thinking and learning processes.

At both colleges, in the end, we were striving for authenticity and alignment across the content, context, and pedagogy of the courses. The students were studying complex systems in the context of several real-world problems and they were using the methods of real scientists and citizens. We wanted everything about the course to reflect how one operates in the real world amid complexity and uncertainty. Of course, we did not fully reach this goal, but the courses and our study of them taught us much about how to craft an "authentic" learning setting for students that requires them to be creative, work across boundaries, and integrate ideas from multiple perspectives.

Evidence of Integrative Science Learning

We draw evidence from both courses over a range of student work that includes responses to integrative writing assignments, weekly data-rich science homework, in-class activities, and final integrative projects. Using qualitative methods of analysis, we analyzed student writing to reveal learning related to our course goals for science (see below) and to "integrative moves" over a range of sophistication. For example, we draw and expand on the work of Jack Mino (2006), who has identified a number of integrative mechanisms observed in student writing. Overall, we analyzed our students' writing in a holistic and inductive way, searching for the most common and shared features and patterns.

From this analysis, we present evidence here that our students are indeed learning in integrative, creative, and scientific dimensions. We show what this learning looks like in multiple courses, institutional contexts, and types of students. To bring the integrative work to life in the context of on-the-ground practice, we provide evidence examples that are detailed and contextual.

With an eye toward the central integrative act of advancing understanding by making connections between disparate ideas and domains of knowledge, we discuss our evidence in two categories: science connections and "stretch." Students were indeed learning the science of complex systems, a science largely new to them that involved systems and connecting concepts from various scientific disciplines. They were also learning science by making fruitful connections across different kinds of boundaries—to other science and nonscience disciplines and to more personal knowledge—with some unexpected surprises. They were also stretching and going beyond what they and we expected. We invited and saw intellectual play in their work, movement into domains of knowledge less familiar to students, and fruitful science learning that emerged from this. All of this was entwined with situations where students integrated ideas, from the course and from their own experiences, to creatively produce understanding that looked "new" to them, and sometimes to us.

The last two sections present one major research finding through a different lens. We distill and abstract common trends, in these particular contexts, for integrative thinking observed across two courses, instructors, and institutions. These trends form the basis for a simple (and yes, reductionist) yet useful framework that helps us think about how integrative student work progresses in sophistication from novice to expert.

Science Connections

As many science instructors do, we organize science learning goals for students into two categories: content (or concepts) and process. For content, we expected students to be able to explain scientists' understanding of the possible causes of climate change on various time scales and to demonstrate a basic understanding of complex systems, including ideas of threshold crossing, feedbacks, nonlinearity, and sensitivity to initial conditions (National Academy of Sciences, 2002). For process, we expected students to understand how climate data were collected and analyzed, and how science knowledge can be both reliable and tentative. We expected them to gain understanding about the iterative process of building science knowledge through the fluid interaction between experimental evidence and theory development.

We have alluded to studying "living," complex, real-world science in our courses. The central scientific focus was on abrupt climate change or the phenomenon of dramatic, tipping point–like shifts in climate over short periods of time (less than a decade). While the evidence for abrupt climate change comes from multiple sources, we focused on a recursive analysis of ice core data from Greenland (Alley, 2002), which provides some of the richest, most detailed data over the past 2 million years. At each pass through climate data further back in time, we discussed the proxies used to collect climate data along with the tentative theories used to explain a sequence of events on earth's surface responsible for fast climate change.

Before we show evidence for students' science learning, it is important to understand what students were connecting as they did this. We designed writing assignments to encourage integrative thought. To some extent, the assignments dictated what the students were connecting. However, analysis of the writing indicated students also made connections we had not invited—or anticipated.

Overall, our analysis reveals five "integrative dimensions" where students learned science through connection making that produced new insights for them. We describe here these five domains of integration (Stewart, 2006), along with evidence from student work that demonstrates science learning at various degrees of sophistication.

Different Science Disciplines. Students integrated knowledge from geology, chemistry, astronomy, oceanography, climatology, and other fields in order to understand the science of abrupt climate change. They learned the science through hands-on guided inquiry lessons, minilectures, data analysis exercises, and reading and discussing Alley's *Two-Mile Time Machine.* They demonstrated their understanding and integrative skills through "What do you think now?" exercises that were repeated periodically, by writing about revised understanding in light of new historical data sets on climate, or through creative assignments that involved teaching others.

Student written responses on the data analysis exercises provided a measure of their ability to interpret complicated graphs in the context of the interdisciplinary field of climatology. They completed some exercises in class in small groups, and addressed questions that arose during their work through whole-class discussion. All groups completed the exercises, and their answers demonstrated good understanding of the data. They were initially puzzled by words such as "regime" and "perturbation," in the context of tipping points, but they were using the terms comfortably in discussion by the end of the class period. In the "How fast can climate change?" exercise, students were confronted with noisy data that did not provide clear beginnings and endings to climate changes. This made them initially reluctant to attempt to answer the "how fast" question. When they understood that they were looking at the "real data" that "real scientists" use to make these decisions, they began to demonstrate a deeper understanding of the concept of data analysis and interpretation. They were able to decide on the criteria they would use to answer the question, and they were able to describe their criteria along with their answer to the "how fast" question.

One assignment at Hope asked students to "teach back," in any format, the science they had learned. Students wrote papers, prepared PowerPoint presentations, designed posters, and in several instances developed a board game or a climate change song. At Carleton, students did short weekly writing assignments focused on the science. All students were to demonstrate their understanding of all the new scientific terminology we had covered and provide clear explanations of the causes of abrupt climate change. While there was evidence of differing amounts of student time and effort put into the assignments and projects, students' explanations of climate change demonstrated mastery of such terms as thermohaline circulation, paleothermometry, the carbon dioxide cycle, and the various mechanisms for climate change at play on different timescales.

Science and Nonscience Disciplines. One integrative writing assignment used at both colleges asked students to creatively connect abrupt climate change, an example of a "tipping point" in the physical world, to the abrupt social change in Gladwell's *The Tipping Point.* Gladwell uses examples from psychology, sociology, and economics to create a framework for understanding social tipping points—abrupt change in human networks. Gladwell's "Law of the Few" suggests key social players are critical to the spread of ideas and trends in society (mavens, connectors, and salesmen). In addition, an idea or fad can be "sticky" so it really takes hold and spreads quickly. The "power of context" concept recognizes that powerful environmental factors come into play as well.

Students had read and discussed *The Tipping Point* in class. They now had to develop analogies between tipping points in climate and tipping points in society that led them to new understanding. Here, analogic reasoning became a path to integrative learning. The assignment probed students' understanding of climate as a complex system while asking them to enhance and explore their understanding through the creative use of analogies. In science, analogies are powerful ways to describe and understand phenomena that are often invisible to the human eye. Often, scientific theories take on names that are really all about an analogy (e.g., planetary model for the atom, etc.). However, appropriate use of analogies is really hard for students because they require good understanding, both about the analogous situation and about the one being studied. As one Hope student wrote, "I have to really understand the mechanisms of abrupt climate change in order to create analogies with social science, although I am certainly not claiming total understanding!" When we described this assignment in class, students expressed initial puzzlement. So we did a bit of coaching. Could they see, for instance, concepts in the science of abrupt climate change analogous to Gladwell's "Law of the Few" or "Power of Context"? Students' initial attempts during class discussion inevitably connected the human to the human, and did not include the science: "Al Gore represents the Law of the Few, because his video reached millions of people." With some gentle nudging and supportive small group discussions, students were eventually able to connect the social science framework to the *physical* science of climate change: "The carbon dioxide molecule is like the Law of the Few. . . . It's a tiny molecule that has enormous impact on climate." In the final papers, many students were able to link Gladwell's idea of "connectors" to something comparable, the ocean conveyor belt, in the earth's climate system: "This great personality and ability to meet people allows them to connect tons of people together through the relationships that the connector has made. Thanks to the ocean conveyer belt, all the continents have a common bond that links their climates." A tougher task for students involved linking Gladwell's idea of the "power of context" to a comparable feature in the climate system. Here we saw some students (not a majority) talk about how factors and causes, on earth's surface and beyond, are intertwined in a complex system: "In the science of climate change, the effects of any single factor will depend very greatly on the context of that factor. The effect of a wobble on the earth's axis will depend on many other factors that influence climate. This illustrates the incredible complexity of the science of climate change." Finally, a handful of students used the "power of context" to talk more metacognitively about their shifting views on climate as they engaged in our recursive process of examining climate data. "Coming into this class, I assumed that we were in a global warming period due to human activity. After looking at various graphs of different historical records of global temperatures, I realized how vital the 'context' or time frame of the graph had on my views of our current climate trend. . . . In fact it appears as though we are in an unusually prolonged period of climatic stability." Overall, we found that starting with Gladwell's concepts of abrupt change in the human domain, where our students live every day, was highly engaging, motivating, and powerful for their science learning. Human experience became a powerful starting point for student-made analogies that produced new scientific insight.

Science Knowledge and Science Values. We explored the role of beliefs and values in constructing new science knowledge through reading about the history of abrupt climate change and reading either Kuhn (1996) or a short paper by Doug Allchin (2004) about science and values. Students learned that scientists' beliefs about climate as something that changes only very slowly limited their ability to "see" the abrupt climate change evidence in front of them (Weart, 2003). They wrote about the need for scientists to be "open-minded" with respect to new data that challenges their long-held beliefs.

Science Knowledge and Personal Values. Students reflected on the ways their personal values shaped their beliefs about the physical world and how those beliefs shaped what they think should be done about climate change. Many students wrote of their belief in the importance of human "stewardship" of natural resources, some saw a divine hand in the geological cycles they learned about, and others invoked the importance of social justice in making decisions about climate change.

Science Knowledge, Personal Beliefs, and Decision Making. Students used their new science knowledge to make predictions about future climate change in different parts of the world. These predictions helped shape their decision about what should be done to reduce human vulnerability to climate change.

Finally, one striking result with regard to connections is that students often made the most progress in understanding complex systems by connecting the course learning to their prior background, values, or experiences. The role of the "student perspective" loomed large and propelled their learning in unexpected ways. This trend was so rampant that it is tempting to suggest that for novices who are not well versed in one (or multiple) disciplines, the "student perspective" can be used in positive ways to practice and enhance integrative learning. Students need not have deep knowledge in two disciplines to integrate in a meaningful and powerful way and thus gain new insight. They can be encouraged to do productive integrative thinking across the course content and their own prior learning—with surprising benefits to learning.

Stretch

Our evidence strongly indicates a creative dimension to connected science learning, where students were stretching beyond the familiar and comfortable to experiment with the intellectual play of connection making. In fact, we were doing the same things as teachers as we designed, revised, and taught our courses. At the heart of integrative learning is the notion that new understanding is born from connecting disparate ideas, requiring intellectually flexible and imaginative acts by the learner. The work is risky, creative, and nonlinear. So we looked carefully at our student work with some specific questions. What were our students creating that is new to them, and what does it really look like? What was the *product* of their creative thinking? What processes and mechanisms were they using to create a new product or way of thinking?

We believe that the creative act of integration involves uncomfortable crossings into new and less familiar domains of knowledge—and the inclination to learn through this kind of exploratory approach. In the long run, we are not only interested in what the students produce or learn that is new to them, we are also interested in understanding their thinking processes. We observed that students were really stretching to take on and exercise some new processes for integrative thinking. We were trying to help them move into new domains of knowledge; we coached them on how to use connected thinking intentionally and fruitfully. So one product of the learning in the course involved moving students along in their ability to be open to and inclined to this kind of exploratory thinking through boundary crossings.

Indeed, we found in student writing places where these novice learners stretched beyond our or their expectations, and beyond the boundaries of a typical introductory science course—to create insight that was new to them. There were several areas in which we also observed creative stretch within the interaction of student groups. We observed students stretching in three big areas, integrated over both our courses and college contexts: the science they were learning, the human time frame in relation to simple systems, and the notion that passive learning of facts defines college science learning (Ferrett, 2006).

The stretch beyond science as a discipline involved pulling in perspectives from social science (human networks), something that the course design clearly supported with the *Tipping Point* readings and integrative assignments. What was surprising was the extent to which students drew on their own perspectives, using knowledge, experience, values, beliefs, and sheer curiosity. The strongest pattern we observed was that students drew extensively on human experience, and in productive ways, to more deeply understand the science. As noted above, they would often discuss a phenomenon in the human domain, and then jump to an analogous situation or idea regarding climate change. One Hope student recognized the power of this kind of thinking beyond the domain of science: "These concepts (Gladwell's) come to life more when they repeatedly are evident in different fields." With regard to stretching beyond the human time frame and simple systems, student perspectives often transformed by considering the massive geological timescales involved with climate change, and complex processes on earth's surface. Again, these courses, as with any good geoscience course, are designed for this kind of shift with the recursive emphasis on data interpretation on deep geological timescales. In addition, student work began to show deeper understanding of essential concepts related to complex systems: feedback loops, thresholds, triggers, inherent uncertainty, and connectedness. Student predictions about future abrupt climate change started out simple, naïve, and certain (though incorrect), and increasingly because complex, informed, and less certain (and more likely). Multicausal explanations appropriately became the norm by the end of the courses.

Perhaps most compelling to us, students moved past the passive learning of science as a set of facts to master. As they cycled through data and theories and read about the history of understanding on abrupt climate change, they wrote about inquiry as a recursive, dynamic, and human process. Their writing included more on the role of conflict, resistance, anomaly, belief, and paradigm shifts. Some students, especially those who read Kuhn's work (1996) at Carleton, wrote eloquently about the role of struggle and resistance in the building of knowledge, and about knowledge as tentative.

Finally, we also saw students engage in more active inquiry as they interpreted climate data and posed integrative questions that opened up avenues for potential inquiry. At Carleton, where the coup emerged (opening story for this chapter), students began to move from individual learning to a kind of group inquiry that became more coordinated and student driven. Students generally became more skeptical and in control of their own learning, critique became more prevalent, and metacognition about their own learning emerged.

With regard to increased agency by students in their own learning, we noticed that a few of our students were writing pieces that went far beyond our instructional prompts, reworking an assignment around a purpose that they defined for themselves. And even fewer of them were then pursuing their purpose with some persistence—exploring and connecting their purpose to the assignment prompt and the work of the course. This was indeed unexpected and highly integrative. These students were not only redefining the assignment, but they were moving into a mode of evaluation and critique that in a few cases took a step toward suggesting reasoned action on a complex societal problem (not part of the assignment!). This move toward what we will call "activism" was impressive. These students moved the farthest from a writing prompt that said "write about how science ideas for abrupt climate change relate to Gladwell's ideas about how ideas and fads move through human social systems." A few students completely owned and shaped their assignment into something that had deep meaning—for them. They had integrated our assignment with theirs to create a piece of writing that had a self-defined and larger purpose, analogic reasoning supported by evidence, integrative thinking in multiple dimensions, and a reasoned call to action. This was clearly on the high end of integrative thinking—and indicated a very intentional approach to integrative learning.

Furthermore, once we noticed student ownership and agency, we returned to look again at student writing, particularly the evidence tied to the analogy assignment discussed above using ideas in *The Tipping Point*. Indeed, we started to observe smaller

Table 11.1. "Integrative Movelets" Obtained by Analyzing Student Work, with Prototype Examples from Student Work and Bloom's Taxonomy (1956) Verbs for Cognitive Development

Integrative Movelet	Prototype Example	Bloom's Taxonomy Verbs (Domain)	Frequency Found in Student Writing
Recognize an analogous situation across a traditional boundary	This is like that. "Tagging."	Name, recognize, identify, relate (knowledge, comprehension)	Ubiquitous
Analogic reasoning by compare or contrast	This is like that because . . . This is not like that because . . . Student may bring in data to support compare or contrast reasoning.	Explain, compare, differentiate, analyze, contrast, transfer (comprehension, analysis)	Very common
Analogic reasoning produces a new idea or understanding	. . . and it means that [new idea]. Student produces an idea new to her or him or a better understanding of a concept.	Create, construct, deduce (synthesis)	Less common
Self-defined purpose or problem posing emerges; Questioning begins that not requested by instructor	I will use "this is like or not like that" to address [student purpose].	Create, construct, devise, propose (application)	Uncommon
Adapts analogic reasoning to problem solving; Active inquiry and exploration begins with data gathering, research, etc.	"This is like that" leads me to investigate [a question, problem solve, deeply explore] . . .	Seek information, solve, analyze (application, analysis, synthesis)	Rare to extremely rare
A call for action is made, defined, and argued by student using analogic reasoning and critique.; Relates to student's values or passions	We must revise our policy on [manufacturing constraints for a consumer good] because . . .	Defend, support, recommend, value, critique (evaluation)	Extremely rare

steps students were taking toward self-defined purpose and related inquiry. In short, a scale of sophistication was starting to emerge for integrative learning in this particular context of analogic thinking. We were starting to distill some sense for how integrative learning developed and became more sophisticated.

A Framework for the Development of Integrative Learning

In Table 11.1 we lay out a framework for integrative thinking that emerged from our student work where crossing boundaries using analogies was at the center of their efforts (the *Tipping Point* integrative assignment). Though our evidence emerges from this very specific context and assignment, we think this frame may have utility beyond this context as we think about levels of sophistication in students' integrative work. In the table, we use the word *movelet* to indicate that we are now looking closely over the novice spectrum for integrative learning, accepting even small steps toward integration as evidence for its development.

Throughout this chapter we have tried to give some evidence of these movelet types. The table also gives a rough sense for how common these movelets were in our student work. Overall, we see student thinking move from simply noting a connection to explaining it more fully and in a balanced way, to greater student agency in defining whether and how this analogic thinking gets used to make a reasoned argument that might even lead to a compelling call for action.

Our research has helped us understand several important things. First, we now have a much clearer picture, within the context of analogic reasoning, of what integrative learning looks like. The table, drawn from our students' writing, even starts to map out a developmental dimension to this kind of learning. Second, we have shown that college students can indeed do integrative work as novices—and even first-year students can do this.

Though this framework emerged from the inductive analysis of our students' work, it turns out to be highly resonant with the research literature on development through the college years. For example, the move to agency and self-authorship and connection to personal values are also seen in the schemas proposed for intellectual development. The move to using reason-based evidence and critique exists in a variety of schemas for intellectual development. The use of flexible, exploratory, and creative thinking ties strongly to the literature on "adaptive expertise" (Bransford et al., 2000). Our research project began with a focus on intellectual development, and it has come full circle to a distilled framework that resonates strongly with that literature. This resonance, coupled with our inductive approach to analysis of student work, is promising. Though we do not understand fully how integrative learning progresses in sophistication, we have made a small step forward—in real college classrooms that vary in context, using analysis of real student work.

Acknowledgments

We would like to acknowledge those who helped us think hard about this work—Mary Huber, Richard Gale, and Whitney Schlegel. We also acknowledge our course students, without whom this adventure in integrative learning and research would not have been possible. We had financial support for this work from Hope and Carleton Colleges (including institutional grants from the Howard Hughes Medical Institute), and the Carnegie Foundation for the Advancement of Teaching (CASTL Carnegie Scholar Program, 2005–2006).

References

Allchin, D. (2004). Values in Science: An Introduction. University of Minnesota. Available at http://www1.umn.edu/ships/ethics/values.htm. Accessed Apr. 25, 2012.

Alley, R.B. (2002). *The Two-Mile Time Machine: Ice Cores, Abrupt Climate Change, and Our Future.* Princeton, NJ: Princeton University Press.

Baxter Magolda, M.B. (2000). *Creating Contexts for Learning and Self-Authorship: Constructive-Developmental Pedagogy.* Nashville: Vanderbilt University Press.

Baxter Magolda, M.B. (2001). *Making Their Own Way: Narratives for Transforming Higher Education for Promoting Self Development.* Sterling, VA: Stylus.

Belenky, M.F., Clinchy, B.M., Goldberger, N.R,, and Tarule, J.M. (1986). *Women's Ways of Knowing: The Development of Self, Voice and Mind.* New York: Basic Books.

Bloom, B.S. (1956). *Taxonomy of Educational Objectives, Handbook I: The Cognitive Domain.* New York: David McKay.

Bransford, J.D., Brown, A.L., and Cocking, R.R. (2000). *How People Learn: Brain, Mind, Experience, and School.* Expanded ed. Washington, DC: National Academy Press.

Diamond, J. (2005). *Collapse: How Societies Choose to Fail or Succeed.* London: Viking Penguin.

Ferrett, T.A. (2006). *First-Year Students "Go Beyond" with Integrative Inquiry into Abrupt Climate Change.* Carnegie Scholar Final Snapshot. Stanford, CA: Carnegie Foundation for the Advancement of Teaching. Available at http://sakai.cfkeep.org/html/snapshot.php?id=86948187730227. Accessed Apr. 25, 2012.

Gardner, H. (2004). *Changing Minds: The Art and Science of Changing Our Own Minds and Other People's Minds.* Boston: Harvard Business School Press.

Gladwell, M. (2002). *The Tipping Point: How Little Things Can Make a Big Difference.* New York: Back Bay Books.

Hope College. (2012). *General Education.* GEMS, General Education Math and Science. Available at http://hope.edu/academic/gened/inter/gems.htm. Accessed Apr. 25, 2012.

Huber, M.T., Brown, C., Hutchings, P., Gale, R., Miller, R., and Breen, M. (2007). *Integrative Learning: Opportunities to Connect.* Public Report of the Integrative Learning Project, Association of American Colleges and Universities and Carnegie Foundation for the Advancement of Teaching. Washington, DC: AAC&U; Stanford, CA: Carnegie Foundation.

Huber, M.T. and Hutchings, P. (2003). *Integrative Learning: Mapping the Terrain.* Carnegie Foundation for the Advancement of Teaching and Association of American Colleges and Universities. Available at http://www.carnegiefoundation.org/LiberalEducation/Mapping_Terrain.pdf. Accessed Nov. 13, 2004.

King, P.M. and Kitchener, K.S. (1994). *Developing Reflective Judgment*. San Francisco: Jossey-Bass.

Kuhn, T.S. (1996). *The Structure of Scientific Revolutions*. 3rd ed. Chicago: University of Chicago Press.

Mino, J. (2006). *The Link Aloud: Making Interdisciplinary Learning Visible and Audible*. Carnegie Academy for the Scholarship of Teaching and Learning. Available at http://www.cfkeep.org/html/snapshot.php?id=29945016959631. Accessed Apr. 25, 2012.

National Academy of Sciences. (2002). *Abrupt Climate Change: Inevitable Surprises*. Committee on Abrupt Climate Change. Washington, DC: National Academy Press. Available at http://books.nap.edu/books/0309074347/html. Accessed Dec. 12, 2012.

Perry, W.G., Jr. (1999). *Forms of Intellectual and Ethical Development in the College Years: A Scheme*. San Francisco: Jossey-Bass. (Originally 1968.)

Shulman, L.S. (2005). Pedagogies of Uncertainty. *Liberal Education*, 91 (2): 18–25. Available at http://www.aacu.org/liberaleducation/le-sp05/le-sp05feature2.cfm. Accessed Dec. 12, 2012.

Stewart, J. (2006). *Integrative Learning in the Sciences: Decision Making at the Intersection of Science Knowledge and Student Beliefs and Values*. Carnegie Scholar Final Snapshot. Stanford, CA: Carnegie Foundation for the Advancement of Teaching. Available at from http://www.cfkeep.org/html/snapshot.php?id=11570887557763. Accessed Apr. 25, 2012.

Weart, S. (2003). The Discovery of Rapid Climate Change, *Physics Today*, 56 (8): 30.

12 Facilitating and Sustaining Interdisciplinary Curricula

From Theory to Practice

Whitney M. Schlegel

"NO DISCIPLINARY VIEWPOINT is inherently or universally true or superior to others" (Haynes, 2002, p. xv). This is where the story of the Human Biology Program begins, with an opportunity for faculty from diverse disciplines to set aside disciplinary hegemony and intellectually engage with one another in developing, implementing, and sustaining an interdisciplinary undergraduate degree program at a public research university and to emerge in partnership with their students from this experience with new ways of thinking.

The founding faculty of the Human Biology Program on Indiana University's flagship Bloomington campus described one of their most important roles in this integrative science program as "infecting others," communicating an evidence-based, student-centered, and scholarly approach to undergraduate education. Paralleling the epidemic metaphor employed in the *Tipping Point* (Gladwell, 2000), faculty and students viewed themselves as contagious vectors of positive change and inspired a multidisciplinary community to engage in the difficult intellectual work of building and sustaining a unique interdisciplinary undergraduate degree program in the life sciences.

In 2003–2004 Indiana University positioned itself to facilitate and offer leadership in research and education for the state's Life Sciences Initiative. At the same time the provost created a competitive funding program, Commitment to Excellence, to provide funds for new programs that integrated the research and teaching missions of the university. The initiative funded a proposal for a comprehensive program in human biology, supplying the stimulus to develop an interdisciplinary undergraduate degree program in human biology that would serve the state by attracting, educating, and retaining a life sciences workforce.

A campus conversation, launched with the support of the university administration, brought together faculty and students to explore the meaning of such a program for the Indiana University–Bloomington campus. A shared vision began to emerge, helping mobilize an interdisciplinary learning community.

The Process of Building an Interdisciplinary Program

> Fostering students' abilities to integrate learning—across courses, over time, and between campus and community life—is one of the most important goals and challenges of higher education.—Association of American Colleges and Universities and Carnegie Foundation for the Advancement of Teaching, Integrative Learning Project.

Initiators of the process intentionally structured campus conversations to challenge faculty and student thinking about interdisciplinarity and to consider educational paradigms that facilitate connections across disciplines. Designers invited internal and external scholars on teaching and learning and interdisciplinary experts to speak with faculty and students. Drawing from the literature and their own experience, these scholars provided timely and essential evidence to inform decisions on curricular structure, pedagogy, and content for the interdisciplinary program.

Participants determined that to be truly interdisciplinary, the curriculum must have a place for integration, a core course series that spans the four year degree curriculum; and furthermore, it must respect what we know about college student development. To maximize the vast disciplinary expertise of the faculty, the program should draw from existing courses and group courses in ways that offer unique multidisciplinary perspectives, explicitly concentrating on fields in which campus scholars excelled. Faculty and administrators also assembled foundation courses, which provide essential quantitative and life science skills and content.

Importantly, while the campus conversation enhanced understanding of interdisciplinarity within a science context and fostered enthusiasm for integrating the natural sciences, the humanities, and social sciences across the curriculum, it was unable to dismantle all barriers. The initiative required conscientious efforts to reach out to departments, schools, and other units on campus, and they remain central to sustaining the program. Program events, such as a weekly coffee hour and a brown-bag informal speaker series, and cosponsorship of invited speakers for departmental seminars have served to draw faculty and students into the interdisciplinary community.

Students were eager to take part in shaping the proposed degree program. They formed a student advisory group during the early campus conversations, and this group went on to become the student government within the program, electing officers, ratifying a constitution, and electing a student member to the program's advisory committee. The students assumed a leadership role in the program, convening student call outs, organizing movie nights, and coordinating unique learning opportunities for students and faculty. Students serve as peer instructors in the program's core courses and mentors and resident experts in area schools through their outreach activities. The student voice has been an important one in the program's intellectual work, visioning, and governance.

Within the faculty discipline and the student major reside elements of identity and ownership nurtured in part by deep traditions and understandings, which extend well beyond the university. Successfully fostering a shared vision of integrated disciplines and engaging faculty and students require uncommon intentionality, communication, and community building, with support from all levels of university leadership.

Essential lessons learned from the process are that building institutional momentum for an integrative science curricula require strategic alignment of university and campus goals and initiatives with state ambitions, campus funding for new initiatives, support of university and school leaders, campus conversation facilitated by internal and external experts, early student involvement in shaping vision and curriculum, shared vision of what is possible, and attention to community building.

The Process of Implementing an Interdisciplinary Program

> Interdisciplinary pedagogy is not synonymous with a single process, set of skills, method, or technique. Instead, it is concerned primarily with fostering in students a sense of self-authorship and a situated, partial and perspectival notion of knowledge that they can use to respond to complex questions, issues, or problems.—Haynes (2002, p. xvi)

The program intentionally aligned its approach to teaching and learning with the habits of mind for inquiry of its faculty, harmonizing its mission with that of the institution. During the first Human Biology Summer Institute, an event to support faculty cohorts in their collaborative innovation, faculty examined inquiry-based and collaborative learning pedagogies and explored a constructivist-developmental approach for supporting transformative learning. This process led to use of a team-based and case- and problem-based pedagogy with content organized in three modules, referred to as the "signature pedagogy" of the program's core; to effectively integrate disciplinary perspectives, two faculty members from disparate disciplines would each teach the core courses. Reflections on the process indicate its effectiveness. A natural science professor said, "I now have an emerging understanding of the power of these pedagogical strategies for enhancing learning. . . . [A] goal for the day was to convince us of the value and power of teams and case studies; I'm convinced." A social science professor said, "Scaffolding allows students to do more sophisticated things than they could do on their own spontaneously. This is exciting as it maps the step-by-step process as well as giving the student a taste of what it is like to produce much better work." Teachers brought authentic assessment strategies (e.g., scientific poster sessions and peer review) to each core course to genuinely engage students in the environments and processes inherent to science. A backward design approach guided all learning goals and teaching practices, along with the question "What do we want students to know and be able to do as a consequence of their experiences in Human Biology?" This approach, coupled with an understanding of where to look for scholarly literature on teaching and learning and how to gain access to campus resources for teaching, has inspired faculty to continue to ask questions about their students' learning and to design classroom assessments with the dual purpose of facilitating and understanding student learning.

Marcia Baxter Magolda's work (1992) served as the principal guidepost for developing the learning goals for the core curriculum, while William Perry's (1970) intellectual and ethical development scheme steered student learning outcomes. Faculty decided that students' capacity for connecting their learning in the core curriculum with their area of concentration and their academic learning with their lives outside academia would be best supported by a longitudinal student electronic portfolio. "The underlying theme [of these processes] is that instructors are bound to coach cognitive development that will support self-authorship and to do otherwise would be unethical," said a natural science professor in the summer institute. In the first summer institute, the faculty established learning goals for the core curriculum, structured them within a four-year developmental framework, and constructed syllabi to guide implementation of the courses (see the Appendix). They presented their work, "Putting Theory and Research into Practice in the Development of an Interdisciplinary Undergraduate Major in Human Biology," at the second annual conference of the International Society for the Scholarship of Teaching and Learning in Vancouver, British Columbia.

The program's electronic portfolio became the work of the program's second faculty cohort and summer institute. The session delineated seven competencies (scientific reasoning and inquiry, collaborative problem solving, integrative synthesis, communication, personal and professional identity, ethical reasoning and action, and civic engagement) using a matrix model with increasing expectations for integration and student development with year of advancement (see Table 12.1). These competencies were informed by the program learning goals, the scholarly work used in developing the program, and the work of national leaders in higher education and science education, including AAC&U Liberal Education and America's Promise and Valid Assessment of Learning in Undergraduate Education (VALUE), National Research Council, National Science Foundation, and campuses participating in the Keck–Project Kaleidoscope (PKAL) Facilitating Interdisciplinary Learning Project (FIDL), among others. This session described students as "disciplinary explorers" and novice thinkers in their first year, "multidisciplinary investigators" in Year 2, "interdisciplinary critics" in Year 3, and expert thinkers and "extradisciplinary advocates" in Year 4. Evaluation of the e-portfolio and its associated competencies was based on the evidence students provided for each competency, including artifacts and the sophistication of their reflections on their own work, as well as level of achievement, presentation, organization, and originality.

Table 12.1. Undergraduate Interdisciplinary Degree Program Learning Outcomes Supported by a Longitudinal Electronic Portfolio

Competency/level	Extradisciplinary Advocate (=expert) Year 4 *Advocate*	Interdisciplinary Critic Year 3 *Connect/ Critique*	Multidisciplinary Investigator Year 2 *Investigate*	Disciplinary Explorer (=novice) Year 1 *Explore*
Scientific reasoning and inquiry	Convey implications of results and apply new knowledge in a community context	Connect scientific issues and methodologies to social, cultural, artistic, and historical contexts to frame a complex human problem	Recognize competing scientific, social, cultural, artistic, and historical perspectives related to human dilemmas	Identify scientific process (develop scientific questions, design and conduct investigations, analyze and interpret data), methodologies, and uncertainty in scientific knowledge as observed in practice and in different modes of scientific writing
Collaborative problem solving	Identify, design, implement, and evaluate strategies to confront problems in community, professional, or academic settings	Articulate and analyze problems, brainstorm, and synthesize cooperative strategies for addressing problems	Examine problems from various perspectives and reflect on past and current cooperative approaches and solutions	Devise collective strategies to advance understanding
Integrative synthesis	Demonstrate growth and development of interests and abilities over length of degree in multiple community contexts	Demonstrate growth and development of interests and abilities across coursework and extracurricular activities	Demonstrate growth and development of interests and abilities both in and outside the area of concentration	Demonstrate growth and development of interests and abilities across the first year core, seminar, and other coursework
Communication	Differentiate and implement communication modes to promote understanding within community, professional, and academic settings	Integrate perspectives from various disciplines with different modes of communication and use peer review to advance understanding	Recognize and evaluate appropriate communication mode(s) for a given situation and perspective	Identify and practice different modes of communication and recognize own strengths and weaknesses as a communicator

Table 12.1. Undergraduate Interdisciplinary Degree Program Learning Outcomes Supported by a Longitudinal Electronic Portfolio (continued)

Competency/level	Extradisciplinary Advocate (=expert) Year 4 *Advocate*	Interdisciplinary Critic Year 3 *Connect/ Critique*	Multidisciplinary Investigator Year 2 *Investigate*	Disciplinary Explorer (=novice) Year 1 *Explore*
Personal and professional identity	Represent oneself to a variety of audiences in the context of career objectives and personal goals	Connect personal and professional goals through engagement in career or community activities	Characterize identity in relation to others, including beyond the major and into the community	Define what you know and what you need to know as a college student
Ethical reasoning and action	Use evidence-based, decision-making process to express values and participate in public discourse about issues of social justice	Analyze and discuss controversial issues using an evidence-based process	Argue respectfully for a position using a variety of perspectives (e.g., scientific, cultural, social, historical, legal)	Identify ethical basis of collaborative work and scientific endeavors
Civic engagement	Demonstrate leadership in community setting	Participate actively in community event/ organization	Investigate connections between academic and community issues and experience community event	Develop awareness of community issues and organization

In the e-portfolio, students showcase their academic growth, illustrate unique personal and extracurricular activities, and "make meaning" of connections in their college experiences. The audience for the e-portfolio shifts each year, and a capstone e-portfolio course guides students in a reflective process that leads to a professional portfolio and a reflective retrospective essay that utilizes the program competencies as its frame.

Visiting scholars provided the ideal opportunity for faculty groups and the interdisciplinary community to write about and share their experiences with program development and implementation. One faculty reflective report before a visit to campus by Baxter Magolda asserts that

> although we have emphasized specific accomplishments, perhaps the most important product of our activities in the Human Biology program has been the deep intellectual and emotional investment we have developed in its signature pedagogy and in our growing learning community. Several of us have integrated elements of

the pedagogy into our disciplinary teaching. This speaks to the importance and transferability of the skills and attitudes we developed as a result of our participation in the program as well as to our commitment to its central tenets. As a group and as individuals, we have devoted substantial time and energy to making Human Biology a success. At the end of the day, we consider Human Biology among the most interesting and engaging places we spend our time.

By supporting faculty from diverse disciplines in humanities, social sciences, cultural studies, and the natural sciences in their understanding of theoretical foundations for teaching and learning and engaging them in evidence-based practices, the program changed their thinking about teaching and learning, a shift that influenced their perceptions of their disciplines and their classrooms. "In the sciences there is a tendency to sort of hide in your laboratory and not really relate what you do to the real world," said a natural science professor in an interview. "Human Biology has given me a way to connect what I know about science to having a larger influence on the world." Engaging faculty in asking questions about their students' learning and seeking evidence to answer these questions parallels what they do naturally as scholars and has an essential place in fostering best practices, especially when working across disciplinary boundaries. The program "showed us the evidence, . . . shared with us the theoretical frameworks, and most importantly, . . . changed our thinking about our classrooms so that we now think of our classrooms as places for experimentation," said an English professor in an interview. Questions faculty have about specific components of a course, pedagogy, assignments, exams, and so forth can be easily added to course evaluations or end-of--term reflections. These assessments help inform curricular decisions and revisions and provide an easy way for faculty to ask questions about the learning environment and to begin making connections between their teaching and student learning.

For example, ample scholarly evidence supports the value of a team-teaching approach in interdisciplinary courses; however, there are significant institutional barriers that cannot be ignored. One early offering of the first-year core course asked students explicitly on end-of-semester evaluations about the value they placed on interdisciplinary team teaching, and 94% of the students responded with a score of 3 or higher on a five–point scale, where 5 demonstrated the greatest value. Student comments included: "I thought it was one of the best aspects of the course!" and "I enjoyed seeing other people's areas of interest." In the same semester, in the second-year core course the evaluation form did not ask students explicitly about interdisciplinary team teaching; however, when asked to comment on their favorite things in the course, they responded: "Making social and biological associations"; "How we had 2 different perspectives—sociologist and biologist"; "I was able to look at a disease in more than just the biology of it. I learned how the social and governmental issues affect disease and outbreaks"; "I loved learning the scientific with the sociological view."

Student e-portfolios are rich with data that can inform understanding of student learning, and designers intentionally constructed the portfolio to make visible student intellectual and ethical development and meaning making. It displays student work from the many different corners of the curriculum, and faculty can ask questions that help them better understand, for example, how the support they provide in the core curriculum influences student work in courses outside the core, and in cocurricular and extracurricular experiences.

One faculty member wanted to know if students were in fact making the expected moves in their thinking from the first year to the second year, specifically from thinking of an authority-centered or dualistic thinker to thinking that demonstrates a position of their own and employs diverse forms of evidence and criteria to support their views. Three quotes from student reflections help illustrate what the faculty member was able to qualitatively glean from student e-portfolio reflections, share with program faculty, and present at an AAC&U conference.

In the first example, a first-year student clearly demonstrates an authority-centered position that places the power of decision making for antidepressant drugs with medical schools and physicians. The student acknowledges dichotomous views, but does not provide evidence to support them or the opinion the student voices: "Medical schools should train physicians to determine who needs medicine, therapy, or both. I acknowledge the argument against an overly medicated population. I picture a science-fiction setting where no one is ever upset. However, I think the American public should trust physicians to only medicate those who need it." In the second case, a first-year student demonstrates an understanding of the multiple factors that complicate the development of obesity; however, all of the factors are biological in nature. The student does not discuss social, cultural, and environmental factors or lifestyle choices. The student is moving away from dualism but does not yet demonstrate sophistication in the use of criteria and evidence that allows for a synthesis of ideas beyond a broad concluding statement: "By definition, obesity results when energy consumption exceeds energy expenditure. Other factors which increase the chances of developing obesity include genes, hormonal imbalances (hyperthyroid), and neurological impulse complications. Along with these contributions that put individuals at high risk, specific medications may cause alterations in weight gain/loss. Thus, there are many contributing factors that have caused so many adults to develop obesity." In the third reflection, a second-year student provides a hypothesis and evidence-based reasoning with references to the literature (not shown). The student discusses cultural and social factors as well as biological factors that contribute to the spread of disease, first situating the evidence historically and then connecting this evidence with general factors in the spread of new disease:

> A plausible hypothesis for Angola's cholera outbreak is that the disease occurs as a result of naturally occurring and human created environmental conditions. Poor sanitation, consumption of contaminated water and food, poor education, and exposure to ocean brackish water are significant contributors to the spread of cholera. In addition, the persistent hot climate provides an environment for V. cholerae

and other types of infectious bacteria to cultivate. This increases the chances of infection. Another contributing factor to the spread of disease is human migration. Increases in population density [cause] stress on the sanitation systems and new diseases can be introduced into the already existing population.

With an understanding of how students are advancing their thinking and how they use the classroom experience, faculty can advise students individually and, importantly, make informed adjustments to their teaching, including the choices of content, context, and learning environment. Faculty asking questions about their students' learning is part of the culture of the program and is supported by a larger campus community in the scholarship of teaching and learning that is helping to develop evidence-based practices.

Scholarly inquiry is inherent to the ways in which faculty members work and is consistent with the mission of and reward structure within the institution. Furthermore, as the campus struggles with ways of understanding student learning and its relationship to institution-wide curricular changes and innovations at the school, department, and course levels, it is increasingly looking to faculty scholarship on teaching to help with its assessment of student learning. Ernest Boyer (1990) wrote, "The degree to which this push for better education is achieved will be determined, in large measure, by the way scholarship is defined and, ultimately rewarded" (p. xiii). Providing seamless ways for faculty to connect their disciplinary scholarship and habits of mind with their interdisciplinary teaching may help to facilitate shifts in the campus's teaching culture and reward structure.

Partnerships and collaborations are a vital part of interdisciplinary endeavors. The campus instructional support center was an essential partner in the development, implementation, and assessment of the program's curriculum. Partnerships with the School of Education were instrumental in programmatic inquiry and assessment and worked simultaneously to bridge the language differences associated with interdisciplinary student learning in different educational settings. Jennifer Eastwood and colleagues have recently described significant work resulting from one of these partnerships (Eastwood et al., 2011).

Using qualitative and quantitative methods, Eastwood et al. examined as part of Eastwood's doctoral dissertation research how learning outcomes and key learning experiences differed between human biology majors and biology majors. Although the researchers found no differences in content knowledge, as determined by the Biology Concept Inventory (Klymkowsky and Garvin-Doxas, 2008), between disciplinary and interdisciplinary life science majors, the study revealed that the blending of social and biological perspectives within the context of authentic problems and issues in an interdisciplinary program enhanced students' reasoning in novel situations and fostered broader consideration of multiple perspectives when seeking solutions to complex problems. When thinking about new problems, human biology majors frequently referenced case studies they had encountered in their core courses. For these students situating science within issues, problems and diverse disciplinary contexts provided a larger repertoire from which to reason.

A student described the value of an interdisciplinary approach to science as "[n]ot simply what is science, but how does it affect people and why is it important?" (Eastwood et al., 2011). In contrast to biology majors, human biology students believed that they could confidently speak to the relevancy of an issue and take a stand on issues that employ scientific understanding. This applicability and informed opinion was, however, accompanied by a belief that they did not possess detailed knowledge of science and would need to seek this information, whereas biology students revealed more confidence in their knowledge of science. On the other hand, the biology majors did not think they were provided the opportunity to explore the broader implications of science. "They (professors) pretty much all focus in their own area," one student said. "They never really connect to each other explicitly." Biology students expressed an appreciation for learning the tools of science; one student thought that introducing the social implications of biological issues could "cause a flare up or something. It's not quite the 70s anymore but you still watch out."

Researchers asked students to talk about the personal outcomes of their majors and the learning environment they experienced. Human biology students expressed confidence in their ability to identify multiple perspectives, doubt statements lacking sufficient evidence, respect different viewpoints, consider the broader outlook, use evidence in arguments, and research all sides of an issue or problem. Biology students discussed achieving a passion for science, valuing science, understanding the nature of experimentation, experiencing methodological and investigative failure, and learning diligence, responsibility, and goal setting.

Human biology majors valued a strong sense of community fostered in the program by team-based work and access to professors, while biology majors thought the number of students in their major limited development of peer and faculty relationships until class size decreased in the junior or senior years. Human biology students saw the importance of team teaching for modeling integration and acknowledged the challenges for professors in melding strong disciplinary identities and perspectives. They identified with both the benefits and challenges of collaborative work and developed an appreciation for the importance of conflict resolution. Biology students noted the importance of their research experiences and distinguished these from the laboratory experiences associated with courses that they thought portrayed laboratory science as "rigid steps."

Students are aware that interdisciplinary learning and majors are not mainstream academe. This can both empower them and give them reason for concern. A human biology student said, "I'm worried that I'm going to encounter people who are not interdisciplinary and people who are very rigid scientists and they are not going to be able to appreciate the sociology" (Eastwood et al., 2011). A senior female human biology student said, "Science began as a discipline to me, and now it is also a way of thinking. My work in the last two years has prepared me to intervene in real controversies and to contribute meaningfully to the crucial work of finding real solutions to real problems." Experimentation is a natural part of the life cycle of any classroom, course, or curriculum and inherent to faculty work and identity, especially at a research university. Evidence of student learning reported for different learning environments and populations of students provided faculty with an evidence-based approach to understanding their students' learning and guided their thinking about how best to implement courses and curricula.

Essential lessons learned while implementing an integrative science curricula include the need for dedicated time for faculty-driven development of curriculum and assessment, e-portfolios for longitudinal assessment of interdisciplinary learning outcomes, awareness of the work of national leaders and scholars, support from campus teaching and learning offices, partnerships with campus education experts, scholarly inquiry by faculty as an important means of realizing assessment aims and alignment with the academic mission, and conference attendance and presentations for gaining knowledge of best practices and connecting with teaching and learning communities.

The Process of Sustaining an Interdisciplinary Program

> Senge writes, "When we lead and learn in interdisciplinary community, we discover our roles over time; we purposefully commit to shared values and goals, and we acknowledge a diversity of viewpoints, perspectives, and backgrounds. When people choose to become members of an interdisciplinary community—in all their various roles—they make a commitment to examine and reexamine what membership in that community entails. Living in a world of permanent white water, in a world where the definitions of what it means to learn are changing, those who claim to undertake change must examine how that change effects [sic] every aspect of the system in which they work. Far from heading over a cliff, interdisciplinary work brings life to an organization and connects it with the enduring spirit of living and learning in community."—Haynes (2002)

Sustaining curricular innovations is by far the most difficult step in realizing interdisciplinarity in higher education. The interdisciplinary program in human biology experienced abundant intellectual space and resounding administrative support during its development in part because it intentionally aligned its mission with the strategic directions of the institution and the state. The program faculty recognized that efforts at all stages of building and sustaining an interdisciplinary program would benefit from sharing knowledge with other institutions. The opportunity to talk about different approaches and hear the experiences of other PKAL FIDL teams struggling with the same central question—"what is the utility ('making a difference') of interdisciplinary programs for our campus and what structures best serve this mission?"—was pivotal in advancing institutionalization of the program.

Designers established early a paradigm for leadership and program operation that valued teamwork and collaborative governance. The faculty crafted governing policies and procedures for the program and sought ways to ensure equitable distribution of leadership across all disciplines contributing to it. A faculty reflection on the program noted, "Members of the cohort generally agree that the collaborative governance process which stands at the heart of the Human Biology Program is worthwhile and beneficial on a number of levels. For example, we believe that it creates room for more voices, more opinions, and more insights—not to mention way more fun." Establishing policies to govern teaching in the program's core curriculum remained elusive in part because initially the competitive campus award generously funded teaching. Later, the variance in departmental culture within the college and changes in administration would hamper the program's efforts to establish memorandums of agreement and institutionalize procedures for teaching in the program. The program has relied on the goodwill of faculty and departments, clearly not a sustainable model.

Through formal and informal bridges, the program connected with campus units and mission. Designers sought internal and external grant funding for program implementation and assessment, and when appropriate, faculty applied research grants to the program. Faculty and student connections with K–12 students encouraged recruitment and provided expertise for classroom and outreach activities. Faculty service on myriad campus committees and involvement in governance of colleges and the broader campus helped inform program decisions and strategic use of resources. Space for the program was ad hoc from the beginning, and the program's permanent location was in constant flux. The program was able to secure unique space in the heart of campus as a direct consequence of the program's commitment to building bridges and connections.

In May 2009 the program graduated its first class, comprising 84% women and 21% underrepresented minorities in science. Enhancing diversity is one of four strategic directions for the campus and the program directly supports this by retaining a high minority (14.4%) enrollment and enrolling unusually high numbers of women (74%), especially when compared with other science majors in the college. The issue of underrepresentation within STEM fields for women and racial and ethnic minority students has become a national concern for our colleges and universities (Committee on Underrepresented Groups and the Expansion of the Science and Engineering Workforce Pipeline, 2010). Studies have found that strengthening academic support and encouraging social activities for students, especially at the departmental level, fosters a sense of belonging and positively influences retention of undergraduate underrepresented students. The culture of the interdisciplinary program in human biology builds community in support of learning and encourages student leadership, mentorship, and ownership within the program; through these best practices, the program has cultivated a diverse and robust interdisciplinary science learning community.

Gaining understanding and support for the program from campus leaders faced a challenge in an unusual turnover in campus leadership that transcended the program's development. Leadership turned over multiple times at all levels, including president, provost, and college dean by the time the program graduated its first class. Later in that same year the program would experience what Julie Klein (2009) calls "a perfect storm of political, ideological, academic, and economic arguments, even in the face of counterevidence" (p. 125), forcing changes in program structure and leadership. The program faculty continue to work to sustain the learner-centered, evidence-based, and community-building best practices; however, the program is again adapting to a new administration, on the heels of adopting significant curricular changes mandated by the previous administration.

We must keep in mind that our students are the true change agents; consequently, how we educate them will be as important as what we teach them, for both will guide how they engage with society. Interdisciplinary programs have in the past battled the necessity to employ nontraditional structures to institutionalize their mission and continue to do so. The Human Biology Program experience provides some insights for facilitating learner-centered, evidence-based, and community-minded practices for mobilizing, implementing, and institionalizing an interdisciplinary STEM program.

Essential lessons learned for sustaining an integrative science curricula include that a program requires affiliation with internal units and external communities, faculty-governed policies and procedures with strong and collective leadership, a sense of community with a shared intellectual agenda, common space, adequate personnel and resources, excellent assessment to provide evidence of effectiveness, support of the administration for the program mission, faculty, and leadership.

This story offers some cautions for sustaining interdisciplinary STEM programs. The interdisciplinary program in human biology has undergone significant change within the past two years that can be best understood in the context of what Stuart Henry (2005) describes as "disciplinary hegemony." Changes to the program, and in particular its core curriculum, have occurred as a result of cuts to the program budget, reduced resources for faculty, and removal of class size caps. Confidence in the core curriculum was shaken by assertions that the interdisciplinary core courses were shallow and traded rigor for confusing learning objectives that appeared to be more skill than content driven, and furthermore, they were costly because of the team teaching and small class sizes. The number of core courses was reduced from four to three and faculty-led seminars have been replaced by discussion sections led by graduate students. The constraints of disciplinary structures, cultures, and traditions are an inherent challenge to all forms of interdisciplinary work. Strong leadership is essential for the persistence of interdisciplinary STEM teaching and learning in higher education; and equally important is the distribution of interdisciplinary program leadership and curricular ownership among a critical mass of diverse disciplinary faculty and students so that the intrinsic culture of experimentation, innovation, and responsiveness to change can continue to be encouraged and mentored.

Acknowledgments

The Bloomington campus of Indiana University and its Human Biology Program benefited from participation in a 30-campus three-year project, Keck/PKAL Facilitating Interdisciplinary Learning, and much of the reflective process necessary for sustaining this curricular endeavor was enhanced by participation in this project.

References

Baxter Magolda, M. (1992). *Knowing and Reasoning in College: Gender-Related Patterns in Students' Intellectual Development.* San Francisco: Jossey-Bass.

Boyer, E. (1990). *Scholarship Reconsidered: Priorities of the Professoriate.* San Francisco: Jossey-Bass and Carnegie Foundation for the Advancement of Teaching.

Committee on Underrepresented Groups and the Expansion of the Science and Engineering Workforce Pipeline. (2010). *Expanding Underrepresented Minority Participation: America's Science and Technology Talent at the Crossroads.* Washington, DC: National Academies Press.

Eastwood, J.L., Schlegel, W.M., and Cook, K.L. (2011). Effects of an Interdisciplinary Program on Students' Reasoning with Socioscientific Issues and Perceptions of Their College Experience. In T.D. Sadler (ed.), *Socio-scientific Issues in the Classroom: Teaching, Learning and Research*, pp. 89–126. New York: Springer.

Gladwell, M. (2000). *The Tipping Point: How Little Things Can Make a Big Difference.* New York: Little Brown.

Haynes, C. (ed). (2002). *Innovations in Interdisciplinary Teaching.* Westport, CT: Oryx Press.

Henry, S. (2005). Disciplinary Hegemony Meets Interdisciplinary Ascendancy: Can Interdisciplinary/Interdisciplinary Studies Survive, and If So, How? *Issue in Integrative Studies*, 23: 1–37.

Klein, J.T. (2009). *Creating Interdisciplinary Campus Cultures: A Model for Strength and Sustainability.* San Francisco: Jossey-Bass

Klymkowsky, M.W. and Garvin-Doxas, K. (2008). Recognizing Student Misconceptions through Ed's Tools and the Biology Concept Inventory. *PLoS Biology*, 6: 14–17.

Perry, W.G., Jr. (1970). *Forms of Intellectual and Ethical Development in the College Years: A Scheme.* New York: Holt, Rinehart, and Winston.

Senge, P.M. (1990). *The Fifth Discipline: The Art and Practice of the Learning Organization.* New York: Doubleday.

Appendix: List of Key Works Used in the Development of the Interdisciplinary Program in Human Biology

Association of American Colleges and Universities. (2002). *Greater Expectations: A New Vision for Learning as a Nation Goes to College.* Washington, DC: Association of American Colleges and Universities.

Baxter Magolda, M. (1999). *Creating Contexts for Learning and Self-authorship: Constructive-Developmental Pedagogy.* Nashville: Vanderbilt University Press.

Belenky, M., Clinchy, C.B., Goldberger, N., and Tarule, J. (1996). *Women's Ways of Knowing: The Development of Self, Voice, and Mind.* 10th anniversary ed. New York: Harper Collins Publishers.

Bransford, J.D., Brown, A.L., and Cocking, R.R. (eds.). (2000). *How People Llearn: Brain, Mind, Experience and School.* Washington, DC: National Academy Press.

Huber, M. and Hutchings, P. (2005). *Integrative Learning: Mapping the Terrain.* Washington, DC: Association of American Colleges and Universities; Stanford, CA: Carnegie Foundation for the Advancement of Teaching.

Kegan, R. (2000). What "Form" Transforms? A Constructive-Developmental Approach to Transformative Learning. In J. Mezirow (ed.), *Learning as Transformation: Critical Perspectives on a Theory in Progress,* pp.35–69. San Francisco: Jossey-Bass.

King, P.M. and Kitchener, K.S. (1994). *Developing Reflective Judgment: Understanding and Promoting Intellectual Growth and Critical Thinking in Adolescents and Adults.* San Francisco: Jossey-Bass.

Klein, J.T. (1996). *Crossing Boundaries: Knowledge, Disciplinarities, and Interdisciplinarities.* Charlottesville: University Press of Virginia.

Mansilla, V.B. (2005). Assessing Student Work at Disciplinary Crossroads. *Change,* 37 (January–February): 14–21.

Mentkowski, M. and Associates. (2000). *Learning That Lasts: Integrating Learning, Development, and Performance in College and Beyond.* San Francisco: Jossey-Bass.

Michaelsen, L.K., Knight, A.B., and Fink, L.D. (2004). *Team-Based Learning: A Transformative Use of Small Groups in College Teaching.* Sterling, VA: Stylus.

National Center for Case Study Teaching in Science. (2008). National Center, SUNY Buffalo. Available at http://sciencecases.lib.buffalo.edu/cs/. Accessed May 18, 2011.

Nelson, C.E. (1989). Skewered on the Unicorn's Horn: The Illusion of Tragic Tradeoff between Content and Critical Thinking in the Teaching of Science. In L. Crow (ed.), *Enhancing Critical Thinking in the Sciences,* pp.17–27. Washington, DC: Society for College Science Teachers.

Perry, W.G. (1998). *Forms of Ethical and Intellectual Development in the College Years: A Scheme.* San Francisco: Jossey-Bass. (Originally 1970.)

Schlegel, W.M. and Pace, D. (2004). Using Collaborative Learning to Decode the Disciplines in Physiology and History. In J. Mittendorf and D. Pace (eds.), *Decoding the Disciplines: How Do We More Effectively Bring Students into the Thinking We Do in Our Disciplines?* New Directions for Teaching and Learning, vol. 98. San Francisco: Jossey-Bass.

Wiggins, G. and McTighe, J. (2000). *Understanding by Design*. Upper Saddle River, NJ: Merrill Prentice Hall.

Contributors

Robert Brooker is Professor of Genetics, Cell Biology, and Development at the University of Minnesota, Twin Cities. He is a life sciences education mentor in the National Academies/HHMI Summer Institutes for Undergraduate Biology Education.

Mike Burke is Professor of Mathematics at the College of San Mateo. His work as a Carnegie Scholar focused on integrative learning, the idea that mathematics, in particular, should not be taught as a completely separate discipline, in isolation from all other disciplines; rather, mathematics offers a powerful way to understand the world and to address many of the problems that arise in other disciplines.

Brett Couch is Instructor in the Department of Botany at the University of British Columbia.

Tricia A. Ferrett is Professor of Chemistry at Carleton College and founder and former director of the Carleton Interdisciplinary Science and Math Initiative.

Matthew A. Fisher is Associate Professor of Chemistry at Saint Vincent College. Involved with the SENCER Project for over a decade and developer of two SENCER model courses, he is also a Senior Fellow with the National Center for Science and Civic Engagement (www.ncsce.net). Fisher also directed Saint Vincent's faculty development program for seven years.

Richard A. Gale is Director of the Institute for Scholarship of Teaching and Learning at Mount Royal University (Calgary, AB, Canada). He was a Senior Scholar at the Carnegie Foundation for the Advancement of Teaching and Director of the Carnegie Academy for the Scholarship of Teaching and Learning (CASTL) Higher Education Program.

David R. Geelan is Senior Lecturer in Science Education at Griffith University, Australia, and author of "Weaving Narrative Nets to Capture Classrooms" and "Undead Theories."

Bettie Higgs is Codirector of Ionad Bairre in the Teaching and Learning Centre at University College Cork, Ireland.

Pat Hutchings is a consulting scholar with the Carnegie Foundation for the Advancement of Teaching, where she was previously Vice President and a Senior Scholar working with the Carnegie Academy for the Scholarship of Teaching and Learning (CASTL).

Gregory Kremer is the Robe Leadership Professor and Chair of the Department of Mechanical Engineering at Ohio University, and founding director of the "Designing to Make a Difference" student project experience.

Xian Liu is Professor of English and has been teaching integrated learning community courses for over ten years at Holyoke Community College, MA.

Kate Maiolatesi is Professor of Sustainable Agriculture and Energy at Holyoke Community College, Chair of the Sustainability Studies Program, and Principal Investigator for two National Science Foundation Clean Energy grants.

David Matthes is Teaching Associate Professor of Biology at the University of Minnesota, Twin Cities.

Jack Mino, Professor of Psychology and what might be called "Interdisciplinary Human Studies," cofounded the Learning Communities Program at Holyoke Community College, MA, and is the current coordinator of the program.

Whitney M. Schlegel is Associate Professor of Biology and founding director of the Human Biology Program at Indiana University.

Joanne L. Stewart is Professor of Chemistry at Hope College.

Deena Wassenberg is Teaching Assistant Professor in the Biology Program at the University of Minnesota, Twin Cities. She was a life sciences education fellow of the National Academies/HHMI Summer Institutes for Undergraduate Biology Education.

Susan Wick is Professor in the Department of Plant Biology at the University of Minnesota, Twin Cities. She is a life sciences education mentor of the National Academies/HHMI Summer Institutes for Undergraduate Biology Education.

Robin Wright is Associate Dean in the College of Biological Sciences and Professor of Genetics, Cell Biology, and Development at the University of Minnesota, Twin Cities. She is an educational mentor and member of the advisory board of the National Academies/HHMI Summer Institutes for Undergraduate Biology Education.

Index

Note: An italic "*t*" following a page number indicates a table; an italic "*n*" indicates an endnote.

Lightning Source UK Ltd.
Milton Keynes UK
UKHW010022050920
369340UK00004B/543